An Introduction to

Numerical Weather Prediction Techniques

Library of Congress Cataloging-in-Publication Data

Krishnamurti, T. N. (Tiruvalam Natarajan), 1932–
 An introduction to numerical weather prediction techniques / T. N. Krishnamurti,
L. Bounoua.
 p. cm.
 Includes bibliographical references and index.
 ISBN 0-8493-8910-0 (alk. paper)
 1. Numerical weather forecasting. I. Bounoua, L. II. Title.
QC996.K75 1995
551.6'34--dc20   95-20906
                                                                  CIP

© 1996 by CRC Press, Inc.

No claim to original U.S. Government works
International Standard Book Number 0-8493-8910-0
Library of Congress Card Number 95-20906
Printed in the United States of America 1 2 3 4 5 6 7 8 9 0
Printed on acid-free paper

Preface

This book on Introduction to Numerical Weather Prediction Techniques was originally written for trainees of the World Meteorological Organization. They were mostly at the level of senior undergraduate students. The same work is here expanded and updated with entirely new data sets, more detailed code explanations and results. The chapters logically cover in sequence the spectrum of numerical weather techniques extending from finite difference methods, dynamical and thermodynamics exercises and finally towards simple prediction models. Each of the chapters is designed to be self-consistent. However, for commodity all subroutines used in the different chapters are gathered in one unique Fortran library.

A diskette provided with the book contains complete software for all the cited exercises. The main code segments as well as sample data are provided to illustrate some examples. The user should, however, be aware that displays generated using the code do not necessarily match those edited in the text. Furthermore, the graphic software is not part of the library. The codes are written in standard Fortran and are designed to run on a range of workstations as well as on personal computers.

This work came through several years of help from scientists and students in our laboratory to whom we are most grateful. Support for this work over the years comes from the funding to our laboratory from the National Science Foundation, the National Aeronautic and Space Agency, Office of Naval Research and the National Oceanic and Atmospheric Administration to whom we are most grateful. The following are some of the colleagues whose names deserve mentioning for this effort: Fred Carr, M. Kanamitsu, Arun Kumar, John Manobianco, Mukut Mathur, John Molinari, Simon Low Nam, Hau lu Pau, Masato Sugi, and K.S. Yap. Dr. Bounoua also wishes to convey his appreciation to Dr. Piers Sellers for his support.

Thanks are also extended to Robin Kaye for her expertise in typing the manuscript.

Table of Contents

Table of Contents

1

Introduction

This is an introductory workbook on numerical weather prediction methodology. It is written for the senior level undergraduates and first year graduate levels in meteorology. The size of this text is deliberately limited to 13 chapters so that the material could be covered in a one semester laboratory course. A synoptic laboratory with terminals for use of each student is desirable for such a course. This book is also suitable for individual scientists who wish to undertake it for self learning.

The text came about from a training manual the senior author wrote for the World Meteorological Organization in 1982 which interested a wide audience of students and scientists from many educational and research centers around the world. The present text has been extensively revised, extended and worked out with newer data sets. A diskette is provided with the text that carries these sample data sets and the codes.

This workbook starts out with an introduction to finite difference methodology. Space differencing techniques include first, second and fourth order schemes, the treatment of Laplacian and Jacobian operations and solutions of Poisson and Helmholtz type equations. An entire section in this chapter is devoted to the description of a wide variety of most popular time differencing schemes encountered in numerical weather prediction. The stability conditions for each individual scheme are also discussed.

Chapter 3 enumerates a number of techniques on the computation of vertical velocity. The vertical velocity is not an observed meteorological variable, and in most cases its estimation involves the calculation of the horizontal wind divergence. Small uncertainties in horizontal wind measurements cause large errors in the estimation of the vertical velocity. Understanding of the methods for the calculation of vertical velocity is an important issue.

Chapter 4 describes two powerful and popular methods of computing the streamfunctions and velocity potential, namely the relaxation and the Fourier transform techniques. It also introduces the wind pressure relationship. Unlike the middle latitudes, where the geostrophic constraint is important, in the tropics one can see the departures from the geostrophic law by exploring a number of what are called "balance" relationships. Here one solves for the pressure given a wind field. This chapter shows how the pressure field is deduced from the linear and the nonlinear balance laws.

Chapter 5, on objective analysis, introduces four methods for data analysis that range from the simple polynomial approach to optimal interpolation. They illustrate how raw data can be analyzed over an array of grid points.

The physical processes are fairly important in the evolution of weather. Chapter 6 introduces basic physical concepts relevant to numerical weather prediction. Basically, the use of moisture variables in meteorology is introduced together with some algorithms describing their computational aspects. Some principles on stability are also introduced.

A simple convective model illustrating the evolution of buoyancy driven dry thermal is provided as an introduction to convective modelling. The complex subject of cumulus parameterization is introduced in Chapter 7. Some of the most common schemes for the determination of rainfall rates resulting from cumulus convection are presented. The chapter also includes a section on large scale condensation.

The planetary boundary layer is an important component that needs to be modeled. In Chapter 8 the best way to model the fluxes of momentum, heat and moisture from the earth's surface (land as well as ocean) is sought. This chapter presents several methods for the calculation of these fluxes. There is a constant flux layer some tens of meters deep next to the earth's surface. The calculation of the surface fluxes as well as their vertical distribution is presented.

Chapter 9 introduces the radiative transfer calculations. The treatment of short and long wave irradiances, the role of clouds, the energy balance at the earth's surface and the issue of the diurnal change are presented. Only an elementary treatment of this important physical process is highlighted.

In Chapter 10 a simple barotropic model is introduced. For tropical applications the streamfunction of the flows is the basic dependent variable and is obtained from the analyzed wind field. This forecast model makes use of the principle of conservation of absolute vorticity. This is generally regarded as a first useful model for learning numerical weather prediction. This model has practical applicability over certain parts of the tropics (e.g., Eastern Atlantic and West Africa).

A second numerical weather prediction model based on the principle of conservation of potential vorticity is presented in Chapter 11. Here the reader is introduced to the first primitive equations model. The forecast of the wind as well as the geopotential height is performed at a single level.

Diagnostic calculation from model output is an important area. This helps in the interpretation of model output. If the forecasts are skillful in simulating a phenomenon such as a storm then the diagnostic studies can tell us something about the lifecycle of a phenomenon. If the forecast is poor, the diagnostic calculations performed on the model output and on the analyzed (observed) fields can provide some reasons for the model's failure. These are essential ingredients for developing the numerical weather prediction capability and are addressed in Chapter 12.

Chapter 13 enumerates some recent satellite and model-based data sets that have relevance to numerical weather prediction.

It is important to note that many of the illustrations shown in this textbook cannot be reproduced without a graphical software. Furthermore, because the tables presented in the text have been edited, they will not be reproduced exactly by the software. The software included inside the text has also been edited for presentation.

The students taking this course should have some background in elementary dynamic, physical and synoptic meteorology. In addition, a good working knowledge of the Fortran language is a requirement. The following are helpful recommended texts.

1. Wallace and Hobbs, 1977: *Atmospheric Science.*

2. Holton, 1992: *An Introduction to Dynamic Meteorology.*

3. Houghton, 1985: *Physical Meteorology.*

4. Nyhoff and Leestma, 1988: *Fortran 77 for Engineers and Scientists.*

2

Finite Differences

In meteorology, the fundamental equations governing the atmospheric circulation appear, in general, to consist of sets of non-linear partial differential equations which do not have analytical solutions and are solved using numerical methods. The most common operations encountered during the solution of these equations are of the type of first and second derivative, Jacobian and Laplacian. These operators are spatial derivatives and require the knowledge of the variable at a fixed time. Time derivatives are also common in numerical weather prediction equations; however, because the future state variable is unknown, the finite difference scheme carries time dependent errors which may amplify during the integration and result in computational unstability. Consequently, time integration of numerical weather prediction equations is carried out using special techniques which are discussed separately in this chapter.

The approximation of spatial derivatives at a given point is based on Taylor's expansion of the variable about that point. The values of the variable are supposedly known at discrete points in space and different combinations of Taylor's expansions may lead to various degrees of accuracy in the estimation of the function's derivatives.

1. Finite Difference Formulation

Assume a function $u(x)$, is known at regularly spaced discrete locations separated by a distance Δx. The derivatives of $u(x)$ can then be obtained using finite differences. Taylor's expansion of $u(x)$ about point x is given by

$$u(x + \Delta x) = u(x) + \frac{du}{dx}\bigg|_x \frac{\Delta x}{1!} + \frac{d^2u}{dx^2}\bigg|_x \frac{\Delta x^2}{2!} + \dots + \frac{d^nu}{dx^n}\bigg|_x \frac{\Delta x^n}{n!} \qquad (2.1)$$

or, if the finite increment Δx is negative,

$$u(x - \Delta x) = u(x) - \frac{du}{dx}\bigg|_x \frac{\Delta x}{1!} + \frac{d^2u}{dx^2}\bigg|_x \frac{\Delta x^2}{2!} + \ldots + (-1)^n \frac{d^nu}{dx^n}\bigg|_x \frac{\Delta x^n}{n!} \quad (2.2)$$

2. First Derivative

From these expansions, three different expressions may be formulated for the estimation of the first derivative of the function u.

$$\frac{du(x)}{dx}\bigg|_x = \frac{u(x+\Delta x) - u(x)}{\Delta x} + \frac{d^2u(x)}{dx^2}\frac{\Delta x}{2!} + \ldots \quad (2.3)$$

or

$$\frac{du(x)}{dx}\bigg|_x = \frac{u(x) - u(x-\Delta x)}{\Delta x} + \frac{d^2u(x)}{dx^2}\frac{\Delta x}{2!} + \ldots \quad (2.4)$$

or finally,

$$\frac{du(x)}{dx}\bigg|_x = \frac{u(x+\Delta x) - u(x-\Delta x)}{2\Delta x} + 2\frac{d^3u(x)}{dx^3}\frac{\Delta x^2}{3!} + \ldots \quad (2.5)$$

The order of accuracy of a numerical scheme is defined as the degree of the largest term neglected in the expansion during the approximation of the function. Therefore, (2.3), (2.4) and (2.5) may be written as

$$\frac{du(x)}{dx}\bigg|_x = \frac{u(x+\Delta x) - u(x)}{\Delta x} + \varepsilon(\Delta x) \quad (2.6)$$

$$\frac{du(x)}{dx}\bigg|_x = \frac{u(x) - u(x-\Delta x)}{\Delta x} + \varepsilon(\Delta x) \quad (2.7)$$

and

$$\frac{du(x)}{dx}\bigg|_x = \frac{u(x+\Delta x) - u(x-\Delta x)}{2\Delta x} + \varepsilon(\Delta x^2) \quad (2.8)$$

where $\varepsilon(\Delta x)$ and $\varepsilon(\Delta x^2)$ represent the errors in the estimation of the derivatives and are called errors of the first and second order of Δx, respectively. Equations (2.6) and (2.7) are also referred to as derivatives of first order accuracy while (2.8) is said to be of the second order accuracy. Because of the stencil of points used in the evaluation of the finite differences, the above schemes are named the forward, backward and centered differences, respectively (Fig. 2.1).

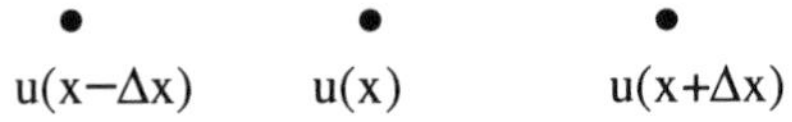

$$\bullet \qquad\qquad \bullet \qquad\qquad \bullet$$
$$u(x-\Delta x) \qquad u(x) \qquad u(x+\Delta x)$$

Figure 2.1: 3 point stencil

The same procedure can be extended to obtain the first derivative of the function to a fourth order accuracy. The fourth order scheme is, of course, more accurate, but requires the knowledge of the values of the function at four adjacent points. This scheme involves the following equations.

$$u(x - 2\Delta x) = u(x) - \frac{du}{dx}\bigg|_x 2\Delta x + \frac{d^2u}{dx^2}\bigg|_x \frac{(2\Delta x)^2}{2!} - \frac{d^3u}{dx^3}\bigg|_x \frac{(2\Delta x)^3}{3!} + \dots$$

$$+ (-1)^n \frac{d^nu}{dx^n}\bigg|_x \frac{(2\Delta x)^n}{n!} \tag{2.9}$$

$$u(x - \Delta x) = u(x) - \frac{du}{dx}\bigg|_x \Delta x + \frac{d^2u}{dx^2}\bigg|_x \frac{\Delta x^2}{2!} - \frac{d^3u}{dx^3}\bigg|_x \frac{\Delta x^3}{3!} + \dots$$

$$+ (-1)^n \frac{d^nu}{dx^n}\bigg|_x \frac{\Delta x^n}{n!} \tag{2.10}$$

$$u(x + \Delta x) = u(x) + \frac{du}{dx}\bigg|_x \Delta x + \frac{d^2u}{dx^2}\bigg|_x \frac{\Delta x^2}{2!} + \frac{d^3u}{dx^3}\bigg|_x \frac{\Delta x^3}{3!} + \dots$$

$$+ \frac{d^nu}{dx^n}\bigg|_x \frac{\Delta x^n}{n!} \tag{2.11}$$

$$u(x + 2\Delta x) = u(x) + \frac{du}{dx}\bigg|_x 2\Delta x + \frac{d^2 u}{dx^2}\bigg|_x \frac{(2\Delta x)^2}{2!} + \frac{d^3 u}{dx^3}\bigg|_x \frac{(2\Delta x)^3}{3!} + \ldots$$

$$+ \frac{d^n u}{dx^n}\bigg|_x \frac{(2\Delta x)^n}{n!} \tag{2.12}$$

The fourth order accurate scheme is formulated as a combination of (2.9) through (2.12) such as the terms in Δx^2, Δx^3 and Δx^4 vanish. This may be obtained by writing

$$\frac{du}{dx}\bigg|_x \Delta x = Au(x) + B[u(x+\Delta x) - u(x-\Delta x)] + C[u(x+2\Delta x) - u(x-2\Delta x)]$$

$$+ \varepsilon(\Delta x^5) \tag{2.13}$$

The terms in brackets may be expanded to give

$$[u(x + \Delta x) - u(x - \Delta x)] = 2\frac{du}{dx}\bigg|_x \Delta x + \frac{d^3 u}{dx^3}\bigg|_x \frac{\Delta x^3}{3} + \frac{d^5 u}{dx^5}\bigg|_x \frac{\Delta x^5}{60} + \ldots$$

$$+ 2\frac{d^{2n+1} u}{dx^{2n+1}}\bigg|_x \frac{\Delta x^{2n+1}}{(2n+1)!} \tag{2.14}$$

and

$$[u(x+2\Delta x)-u(x-2\Delta x)] = 2\frac{du}{dx}\bigg|_x 2\Delta x + \frac{d^3 u}{dx^3}\bigg|_x \frac{(2\Delta x)^3}{3} + \frac{d^5 u}{dx^5}\bigg|_x \frac{(2\Delta x)^5}{60} + \ldots$$

$$+ 2\frac{d^{2n+1} u}{dx^{2n+1}}\bigg|_x \frac{(2\Delta x)^{2n+1}}{(2n + 1)!} \tag{2.15}$$

From (2.13), (2.14) and (2.15) one obtains

$$\frac{du}{dx}\bigg|_x \Delta x = Au(x) + (2B + 4C)\frac{du}{dx}\bigg|_x \Delta x + (B + 8C)\frac{d^3 u}{dx^3}\bigg|_x \frac{\Delta x^3}{3} + (\Delta x^5) \tag{2.16}$$

where the coefficients A, B, C are obtained as solutions of the following system:

$$\left\{ \begin{array}{l} A = 0 \\ 2B + 4C = 1 \\ B + 8C = 0 \end{array} \right. \qquad (2.17)$$

Therefore, the fourth order accurate estimate of the first derivative of the function may take the following final form,

$$\left. \frac{du}{dx} \right|_x = \frac{4}{3} \left[\frac{u(x+\Delta x) - u(x-\Delta x)}{2\Delta x} \right] - \frac{1}{3} \left[\frac{u(x+2\Delta x) - u(x-2\Delta x)}{4\Delta x} \right] \qquad (2.18)$$

3. Second Derivative

The second order accurate second derivative of $u(x)$ may be easily obtained by adding (2.10) and (2.11). That is,

$$u(x + \Delta x) + u(x - \Delta x) = 2u(x) + 2 \left. \frac{d^2 u}{dx^2} \right|_x \frac{\Delta x^2}{2!} + 2 \left. \frac{d^4 u}{dx^4} \right|_x \frac{\Delta x^4}{4!} + \dots$$

$$+ 2 \left. \frac{d^{2n} u}{dx^{2n}} \right|_x \frac{\Delta x^{2n}}{2n!} \qquad (2.19)$$

Therefore,

$$\left. \frac{d^2 u}{dx^2} \right|_x = \frac{u(x+\Delta x) - u(x-\Delta x) - 2u(x)}{\Delta x^2} + \varepsilon(\Delta x^2) \qquad (2.20)$$

Adding (2.9) and (2.12) and substituting for $\left. \frac{d^4 u}{dx^4} \right|_x \frac{\Delta x^4}{4!}$ from (2.19), the fourth order accurate second derivative is expressed as

$$\frac{d^2u}{dx^2} = \frac{1}{\Delta x^2} \left[-\frac{5}{2} u(x) + \frac{4}{3} \{u(x + \Delta x) - u(x - \Delta x)\} - \frac{1}{12} \{u(x + 2\Delta x) - u(x - 2\Delta x)\} \right] + \varepsilon(\Delta x^4)$$

(2.21)

Although for many diagnostic studies the second order accurate scheme appears adequate, fourth order schemes are preferred in numerical weather prediction. Two subroutines *DDX2* and *DDX4* estimating the first derivative at the second and fourth order accuracy, respectively, together with a driver program (*DERIV*), are provided. Results from this simple exercise are summarized in Table 2.1.

Table 2.1: Second and Fourth order derivatives of $f(z) = p_0 \exp(-az)$

Analytical solution	Second order estimate	Fourth order estimate
-0.125062	-0.117558	
-0.110360	-0.110648	
-0.097386	-0.097640	-0.097386
-0.085938	-0.086162	-0.085937
-0.075835	-0.076033	-0.075834
-0.066920	-0.067095	-0.066920
-0.059053	-0.059207	-0.059053
-0.052111	-0.052247	-0.052111
-0.045985	-0.046105	
-0.040579	-0.043226	

```
        program DERIV
c
c    this simple program computes the first derivative of a function
c    using the second and fourth order accurate schemes. the following
c    driver estimates the derivatives of  p(z) = po*exp(-a*z)
c
        parameter(l=10)
```

```
          real z(1),p(1),p2(1),p4(1),anal(1)
c
          data z /0000.,1000.,2000.,3000.,4000.,
     &              5000.,6000.,7000.,8000.,9000./
          data a,p0,dz/0.000125062,1000.,1000./
c
c     initialize the work arrays
c
          do 2100 k = 1, 1
             p2(k)  = 0.
             p4(k)  = 0.
 2100   continue
c
          do 2102 k = 1, 1
c
c     construct the function p(z)
c
          p(k) = p0*exp(-a*z(k))
c
c     compute the analytical derivative
c
          anal(k) = -a*p0*exp(-a*z(k))
c
 2102   continue
c
c     compute the second order estimate
c
          call DDX2 (p,p2,l,dz,1)
c
c     compute the fourth order estimate
c
          call DDX4 (p,p4,l,dz)
c
c     write output. The 4th order derivative is omitted at the first
c     and last 2 points since it is not defined as 4th order.
c
          write (6,1000)
          write (6,1001)
c
          do 2104 k = 1, 1
             if (k.le.2.or.k.ge.(l-1)) then
          write(6,1002) anal(k), p2(k)
                else
```

```
        write(6,1003) anal(k), p2(k), p4(k)
            endif
 2104   continue
c
 1000   format (5x,'analytical',15x,'second order',13x,'fourth order')
 1001   format (5x,' solution ',15x,' estimate ',13x,' estimate ',/)
 1002   format (5x,f10.6,15x,f10.6)
 1003   format (5x,f10.6,15x,f10.6,15x,f10.6)
        stop
        end
```

4. The Laplacian Operator

The Laplacian of a function u(x,y) is formally defined as

$$\nabla^2 u(x,y) = \frac{\partial^2 u}{\partial x^2} + \frac{\partial^2 u}{\partial y^2} \tag{2.22}$$

and appears in many diagnostic and prognostic equations in meteorology. The application of its finite analog is found to be very practical for the solution of many problems. The development of the finite difference form of the Laplacian operator is based on the two dimensional Taylor's expansion about a point (a,b).

$$u(x,y) = u(a,b) + (x - a)\frac{\partial u(a,b)}{\partial x} + (y - b)\frac{\partial u(a,b)}{\partial y} + \frac{(x-a)^2}{2!}\frac{\partial^2 u(a,b)}{\partial x^2}$$

$$+ \frac{(y-b)^2}{2!}\frac{\partial^2 u(a,b)}{\partial y^2} + (x - a)(y - b)\frac{\partial^2 u(a,b)}{\partial x \partial y} + \ldots \tag{2.23}$$

Assuming a regular grid separation in both x and y directions, Taylor's expansion of the functions u(x±h, y±h) about (x,y) may be expressed as

$$u(x+h,y+h) = u(x,y) + h(\frac{\partial}{\partial x} + \frac{\partial}{\partial y}) u(x,y) + \frac{h^2}{2!}(\frac{\partial}{\partial x} + \frac{\partial}{\partial y})^2 u(x,y) + \ldots \tag{2.24}$$

$$u(x-h,y-h) = u(x,y) - h(\frac{\partial}{\partial x} + \frac{\partial}{\partial y}) u(x,y) + \frac{h^2}{2!}(\frac{\partial}{\partial x} + \frac{\partial}{\partial y})^2 u(x,y) + \dots \quad (2.25)$$

$$u(x-h,y+h) = u(x,y) - h(\frac{\partial}{\partial x} - \frac{\partial}{\partial y}) u(x,y) + \frac{h^2}{2!}(\frac{\partial}{\partial x} - \frac{\partial}{\partial y})^2 u(x,y) + \dots \quad (2.26)$$

$$u(x+h,y-h) = u(x,y) + h(\frac{\partial}{\partial x} - \frac{\partial}{\partial y}) u(x,y) + \frac{h^2}{2!}(\frac{\partial}{\partial x} - \frac{\partial}{\partial y})^2 u(x,y) + \dots \quad (2.27)$$

where h may be regarded as the distance between two consecutive grid points. Similarly four other equations can be written for $u(x,y+h)$, $u(x-h,y)$, $u(x,y-h)$ and $u(x+h,y)$ about (x,y). Adding (2.24) through (2.27) the following expression is obtained,

$$u(x+h,y+h) + u(x-h,y+h) + u(x-h,y-h) + u(x+h,y-h) = 4u(x,y)$$

$$+ 2h^2 (\frac{\partial^2 u}{\partial x^2} + \frac{\partial^2 u}{\partial y^2}) + \frac{h^4}{6} (\nabla^4 u + 4 \frac{\partial^4 u}{\partial x^2 \partial y^2})$$

$$+ \frac{h^6}{180} (\nabla^6 u + 12 \nabla^2 \frac{\partial^4 u}{\partial x^2 \partial y^2}) + \varepsilon(h^8) \quad (2.28)$$

Using the other four expansions one may form the sum,

$$u(x,y+h) + u(x-h,y) + u(x,y-h) + u(x+h,y) = 4u(x,y) + h^2 (\frac{\partial^2 u}{\partial x^2} + \frac{\partial^2 u}{\partial y^2})$$

$$+ \frac{h^4}{12} (\frac{\partial^4 u}{\partial x^4} + \frac{\partial^4 u}{\partial y^4}) + \frac{h^6}{360} (\frac{\partial^6 u}{\partial x^6} + \frac{\partial^6 u}{\partial y^6}) + \varepsilon(h^8) \quad (2.29)$$

which leads to the 5 point stencil second order Laplacian (Fig. 2.2)

$$\nabla^2 u = \frac{1}{h^2} [u(x,y+h) + u(x-h,y) + u(x,y-h) + u(x+h,y) - 4u(x,y)] + \varepsilon(h^2)$$

$$(2.30)$$

Similarly, the use of (2.28) and (2.29) produces a second order Laplacian on a 9 point stencil (Fig. 2.3).

$$\nabla^2 u = \frac{1}{6h^2} [4\{u(x+h,y) + u(x-h,y) + u(x,y+h) + u(x,y-h)\} + \{u(x+h,y+h)$$

$$+ u(x-h,y+h) + u(x+h,y-h) + u(x-h,y-h) - 20u(x,y)] + \varepsilon(h^2) \quad (2.31)$$

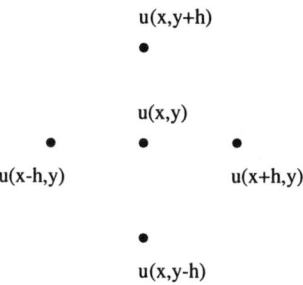

Figure 2.2: 5 Point Stencil

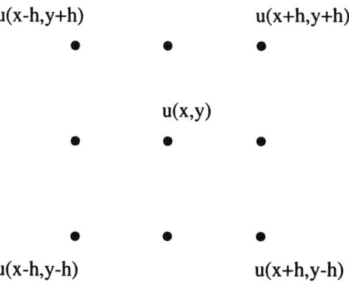

Figure 2.3: 9 Point Stencil

It should be noted that both (2.30) and (2.31) are second order accurate schemes. The nine point stencil is, in general, more accurate than the five point stencil scheme for many applications. Nevertheless, in the solution of the Laplace equation for inhomogeneous boundary conditions the Laplacian is identically zero, making the 9 point stencil scheme a fourth order accurate. A general fourth order accurate 9 point stencil Laplacian solver may be obtained through an iterative technique.

The procedure consists of evaluating successively $\nabla^4 g = \nabla^2 f$, where $f = \nabla^2 g$, using the 9 point stencil on the iterated magnitude of f starting from f=0 on the first iteration.

A computer code computing these different Laplacians (*LAPLACIAN*) is presented. The data used to evaluate the different estimates of the Laplacian are generated on line by a trigonometric function whose analytic solution is known. The root mean square errors between the finite difference estimates and the theoretical solution are also calculated. The driver uses three different subroutines, *LAP94*, *LAP92* and *LAP52*, which estimate the 9 point fourth order, the 9 point second order and the 5 point second order Laplacians, respectively. Outputs illustrating this example are shown in Table 2.2.

Table 2.2: Computation of different order accurate Laplacians

9 pt. 4th order	RMS = 0.3526635E-07	RMS ERROR/RMS ZTA = 0.195E+01 %
9 pt. 2nd order	RMS = 0.2346851E-06	RMS ERROR/RMS ZTA = 0.129E+02 %
5 pt. 2nd order	RMS = 0.1656866E-06	RMS ERROR/RMS ZTA = 0.914E+01 %

Point I	J	Analytical solution	9 pt. 4th order estimated solution	9 pt. 2nd order estimated solution	5 pt. 2nd order estimated solution
1	4	0.30498759E-05	0.30498759E-05	0.30498759E-05	0.30498759E-05
2	4	-0.41585830E-04	-0.41345917E-04	-0.34947032E-04	-0.36936442E-04
3	4	-0.24536503E-04	-0.23390865E-04	-0.20471325E-04	-0.21700844E-04
4	4	0.30636264E-04	0.31358573E-04	0.26373036E-04	0.27602553E-04
5	4	0.47685582E-04	0.47303711E-04	0.40848721E-04	0.42838128E-04
6	4	0.30498677E-05	0.20453918E-05	0.29508433E-05	0.29508442E-05
7	4	-0.41585830E-04	-0.41290550E-04	-0.34947028E-04	-0.36936435E-04
8	4	-0.24536515E-04	-0.23376351E-04	-0.20471340E-04	-0.21700860E-04
9	4	0.30636271E-04	0.31288360E-04	0.26373045E-04	0.27602566E-04
10	4	0.47685582E-04	0.47685582E-04	0.47688582E-04	0.47685582E-04

program LAPLACIAN

```
c
c       this program computes the laplacian using the five-point second
c       order,nine-point second order and the iterated nine-point fourth
c       order laplacian schemes.it also computes the root mean square
c       errors and compares the accuracy of the different schemes to the
c       analytical solution.
c
        parameter (l=10,m=20)
c
c       declare variables and defines some constants.
c
        real psi(l,m), zta(l,m), a(l,m)
        real b (l,m), c(l,m), x(l), y(m)
        pi = 4.*atan(1.0)
        h  = 200.
        yk = 2.*pi/1000.
        yl = pi / 1000.
c
        x(1) = 0.
        y(1) = 0.
        do 2200 i = 2, l
        im1 = i-1
        do 2200 j = 2, m
        jm1 = j-1
        x(i) = x(im1) + h
        y(j) = y(jm1) + h
 2200   continue
        sum = 0.
c
c       construct the stramfunction(psi) psi = sin (kx)*sin(ly)+ cos(ly)
c       and the vorticity (zta) as         zta = d2(psi)/dx2 + d2(psi)/dy2
c
        do 2202 i = 1, l
        do 2202 j = 1, m
        psi(i,j)    = sin(yk*x(i)) * sin(yl*y(j)) + cos(yl*y(j))
        zta(i,j)    = -(yk**2+yl**2)*sin(yk*x(i)) * sin(yl*y(j))
     &                    -yl**2 * cos(yl*y(j))
        a(1,j)    = zta(1,j)
        a(l,j)    = zta(l,j)
        a(i,1)    = zta(i,1)
        a(i,m)    = zta(i,m)
        sum = sum + (zta(i,j) / (l*m))**2
```

```
 2202    continue
         su = sqrt( sum )
         n = 1
   25    go to ( 30,40,50,60 ) n
   30    write(6,1000)
c
c    compute the 9 pts 4th order.
c
         call LAP94 (psi,a,b,c,h,l,m)
c
         go to 70
   40    write(6,1001)
c
c    compute the 9 pts 2th order.
c
         call LAP92 (psi,a,h,l,m)
c
         go to 70
   50    write(6,1002)
c
c    compute the 5 pts 2th order.
c
         call LAP52 (psi,a,h,l,m)
c
   70    continue
         dif = 0.
         do 2204 i = 1, l
         do 2204 j = 1, m
         dif = dif + ((zta(i,j) - a(i,j)) / (l*m))**2
 2204    continue
         dif = sqrt(dif)
         sum = (dif / su ) * 100
         write(6,1003) dif, sum
c
c    output display for one colone.
c
         write(6,1004) ((i,j,zta(i,j),a(i,j),i=1,l), j=4,4)
         n = n + 1
         go to 25
   60    continue
 1000    format(//,20x,'nine points fourth order laplacian scheme.'//)
 1001    format(//,20x,'nine points second order laplacian scheme.'//)
 1002    format(//,20x,'five points second order laplacian scheme.'//)
```

```
1003   format(9x, 'rms error = ',e13.7,8x, 'rms error/rms zta = ',
      &   e9.3,1x, 'percent.'//,2x,'i  j',5x, 'analytical sol',1x,'estimated
      &   sol.',2x,'i  j',5x, 'analytical sol',1x,'estimated sol',/)
1004   format( 2(2i3,4x,2e15.8) )
       stop
       end
```

5. The Jacobian Operator

Jacobian is also a common operator encountered during the solution of many geophysical problems. It appears mostly in the non-linear advective terms. For example, in the vorticity equation, the advection of vorticity by the horizontal wind is given by

$$A_d = - \vec{V}_g . \nabla_p \xi \qquad (2.32)$$

$\vec{V}_g$ is the geostrophic wind defined as

$$\vec{V}_g = \vec{k} \times \nabla \left[\frac{gz}{f_o} \right] = \vec{k} \times \nabla \psi \qquad (2.33)$$

where g is the acceleration of the gravity, z is the height and f_o is the Coriolis parameter. The relative vorticity, ξ, is defined as

$$\xi = \vec{k} . \nabla \times \vec{V} = \nabla^2 \psi \qquad (2.34)$$

where ψ is the geostrophic streamfunction. Thus,

$$A_d = - \vec{k} \times \nabla \psi . \nabla_p \xi = \frac{\partial \psi}{\partial y} \frac{\partial \xi}{\partial x} - \frac{\partial \psi}{\partial x} \frac{\partial \xi}{\partial y} \qquad (2.35)$$

which in symbolic form can be written as

$$A_d = - J(\psi, \xi) \qquad (2.36)$$

J being the Jacobian. This operator appears in many equations in which some quantities are invariant. Therefore, when a finite difference analog is applied to such equations, caution should be made that the errors introduced by the difference method will not alter the conservation principles. For example, in the barotropic dynamic, the Jacobian appears in the vorticity equation as

$$\frac{d\xi_a}{dt} = - J(\psi,\xi) - \beta \frac{\partial\psi}{\partial x} \tag{2.37}$$

where $\beta = \dfrac{\partial f}{\partial y}$, and f is the Coriolis parameter. The variables ξ_a and ξ represent the absolute and relative vorticity, respectively. When integrated over a closed domain, this equation has two important domain invariant quantities. These are the mean total kinetic energy $\overline{\left[\dfrac{\nabla^2 \psi}{2}\right]^2}$ and the mean square vorticity $\overline{(\nabla^2\psi + f)^2}$. These integral invariants are discussed thoroughly in Chapters 10 and 11. Much has been done by Arakawa (1966) on the construction of Jacobian finite difference analogs. Three forms were proposed,

i. $$J(\psi,\xi) = \frac{\partial\psi}{\partial x}\frac{\partial\xi}{\partial y} - \frac{\partial\psi}{\partial y}\frac{\partial\xi}{\partial x} \tag{2.38}$$

ii. $$J(\psi,\xi) = \frac{\partial}{\partial x}\left(\frac{\partial\xi}{\partial y}\right) - \frac{\partial}{\partial y}\left(\frac{\partial\xi}{\partial x}\right) \tag{2.39}$$

iii. $$J(\psi,\xi) = \frac{\partial}{\partial y}\left(\xi\frac{\partial\psi}{\partial x}\right) - \frac{\partial}{\partial x}\left(\xi\frac{\partial\psi}{\partial y}\right) \tag{2.40}$$

The first of these forms is called the advective form while the last two are named the flux forms of the Jacobian. It can be shown that the conservation of the domain averaged kinetic energy, and the mean square vorticity conditions may be expressed respectively as

$$\overline{\psi\ J(\psi,\xi)} = 0 \tag{2.41}$$

and

$$\overline{\xi\ J(\psi,\xi)} = 0 \tag{2.42}$$

5.1 Second Order Jacobian

In this case, the design of the finite difference Jacobian should be adequately chosen to satisfy these constraints. The difference forms of (2.38), (2.39), (2.40) are given by

$$JJ_1 = \frac{1}{4h^2} \left[(\psi_{i+1,j} - \psi_{i-1,j}) \cdot (\xi_{i,j+1} - \xi_{i,j-1}) - (\psi_{i,j+1} - \psi_{i,j-1}) \right.$$

$$\left. \cdot (\xi_{i+1,j} - \xi_{i-1,j}) \right] \qquad\qquad (2.43)$$

$$JJ_2 = -\frac{1}{4h^2} \left[\psi_{i+1,j} (\xi_{i+1,j+1} - \xi_{i-1,j+1}) - \psi_{i-1,j} (\xi_{i+1,j-1} - \xi_{i-1,j-1}) \right.$$

$$\left. - \psi_{i+1,j} (\xi_{i+1,j+1} - \xi_{i+1,j-1}) - \psi_{i-1,j} (\xi_{i-1,j+1} - \xi_{i-1,j-1}) \right] \qquad (2.44)$$

$$JJ_3 = -\frac{1}{4h^2} \left[\xi_{i+1,j} (\psi_{i+1,j+1} - \psi_{i+1,j-1}) - \xi_{i-1,j} (\psi_{i-1,j+1} - \psi_{i-1,j-1}) \right.$$

$$\left. - \xi_{i,j+1} (\psi_{i+1,j+1} - \psi_{i-1,j+1}) - \xi_{i,j-1} (\psi_{i+1,j-1} - \psi_{i-1,j-1}) \right] \qquad (2.45)$$

Arakawa showed that the following second order accurate Jacobian,

$$JJ = \frac{1}{3} (JJ_1 + JJ_2 + JJ_3) \qquad\qquad (2.46)$$

satisfies the integral constraints on the total kinetic energy and the mean square vorticity. To show that, multiply JJ by ψ or ξ and add on a 9 point stencil. Proper cancellations occur between the neighboring points so that the quantities vanish over the entire domain.

Subroutine *JAC* estimating the so-called second order Arakawa Jacobian and satisfying the integral relations of conservation of the total kinetic energy and the mean square vorticity is provided. An example of calculation of the Arakawa Jacobian using the same analytical function as in (*LAPLACIAN*) is performed by program (*JACOBIAN*). Outputs from this driver are presented in Table 2.3.

Table 2.3: Arakawa Jacobian Scheme

Point		
I	J	Estimated Jacobian
1	4	−0.38074524E−09
2	4	−0.14543172E−09
3	4	0.38074521E−09
4	4	0.38074513E−09
5	4	−0.14543179E−09
6	4	−0.47062687E−09
7	4	−0.14543169E−09
8	4	0.38074519E−09
9	4	0.38074519E−09
10	4	−0.38074521E−09

5.2 Fourth Order Jacobian

The fourth order accurate Arakawa Jacobian may be obtained by an adequate combination of 5 second order Jacobians constructed on a 13 point stencil as shown in Figure (2.4). The principle is that the combination would provide an exact cancellation of the second and third order terms. The construction of the fourth order accurate Jacobian is similar to that of the second order and is not developed in this book. The major difficulty encountered during the numerical solution of both the second and fourth order Jacobians resides in the definition of the boundary conditions. The conservation of quadratic invariants, in the finite difference representations, entails the cancellation of $J(\psi,\zeta)$, $\psi J(\psi,\zeta)$ and $\zeta J(\psi,\zeta)$. That is a cancellation of each of the terms in the expansion of the above expression for each grid point with respect to the contributions from its immediate neighbors. This is easy to implement in the interior of the domain but requires an approximation to the finite difference expressions at the boundaries in order to assure the above cancellations. The rationale for this can be found in Arakawa (1966). However, experience showed that a choice of time invariant streamfunction and vorticity conditions along the meridional boundaries and a cyclic boundary condition along the zonal direction is simpler and not detrimental to the quadratic invariants for the 9 point second order Arakawa Jacobian. It was found that these quantities remain nearly constant for 3 to 4 day integrations.

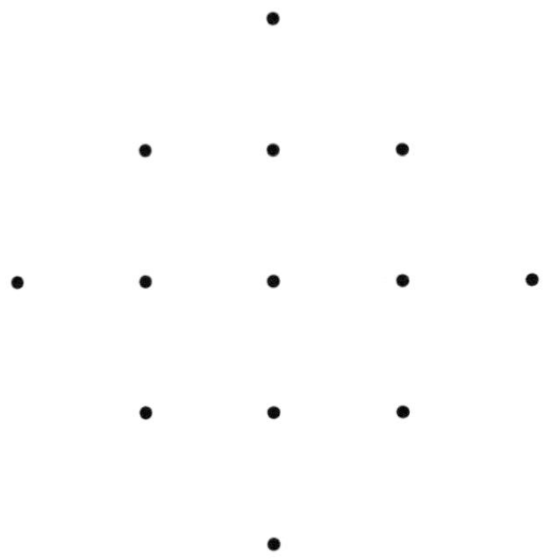

Figure 2.4: 13 point stencil

program *JACOBIAN*

```
c
c     this program computes the arakawa jacobian over a domain of grid
c     -ed data. It uses a nine point fouth order scheme.
c
        parameter (l=10,m=20,l1=l-1,m1=m-1,l2=l-2,m2=m-2)
        real psi(l,m),zta(l,m),a(l,m),x(l),y(m),dx(m)
        pi          = 4.*atan(1.0)
        h           = 200.
        dy          = h
        do 2300 j = 1, m
2300        dx(j) = dy
        yk          = 2.*pi/1000.
        yl          = pi  / 1000.
        x(1)        = 0.
        y(1)        = 0.
        do 2302 i = 2, l
            im1    = i-1
        do 2302 j = 2, m
            jm1    = j-1
        x(i)        = x(im1) + h
        y(j)        = y(jm1) + h
2302    continue
```

```
          sum       = 0.
          do 2304 i = 1, l
          do 2304 j = 1, m
c
c    define the analytical functions psi and zta.
c
          psi(i,j) = sin(yk*x(i)) * sin(yl*y(j)) + cos(yl*y(j))
          zta(i,j) = -(yk**2+yl**2)*sin(yk*x(i))*sin(yl*y(j))-
     &              yl**2 * cos(yl*y(j))
          a(1,j)  = zta(1,j)
          a(l,j)  = zta(l,j)
          a(i,1)  = zta(i,1)
          a(i,m)  = zta(i,m)
          sum     = sum + (zta(i,j) / (l*m))**2
 2304     continue
c
c    compute the jacobian
c
          call JAC (a,psi,zta,dx,dy,l,m,l1,m1,l2,m2)
c
c    display outputs for 1 column.
c
          write (6,1000)
          write (6,1001)
          write (6,1002)((i,j,a(i,j),i=1,l),j=4,4)
 1000     format(//,20x,'arakawa jacobian scheme.',//)
 1001     format(2x,'i  j',10x, 'estimated jacobian',//)
 1002     format( (2i3,8x,e15.8) )
          stop
          end
```

6. Time Differencing

Another problem encountered when solving the equations governing the atmospheric motion is time integration. The detailed mathematical concepts used in finite difference techniques for the solution of partial differential equations is out of the scope of this workbook; however, a general view of some important aspects inherent to the solution of time dependent equations using numerical methods is presented in this section. Unlike space differencing schemes, time differencing schemes seem to meet the accuracy requirements at the first and second order. Higher order schemes appear rather cumbersome and are not of wide use in numerical weather prediction.

For simplicity and without compromising on the generality of the time differencing schemes, the following discussion will focus on the integration of a simple linear equation of the form,

$$\frac{\partial u(x,t)}{\partial t} + c \frac{\partial u(x,t)}{\partial x} = 0 \tag{2.47}$$

where c is a constant. Assume u(x,t) to be a function of the form

$$u(x,t) = Re[u(t) \, e^{ikx}] \tag{2.48}$$

where k is a constant. Substituting in (2.47) and using $\omega = -kc$ yields

$$\frac{\partial u(x,t)}{\partial t} - i\omega \, u(x,t) = 0 \tag{2.49}$$

Integrating between two time levels, t_0 and t, and assuming $t_0 = 0$ gives

$$u(x,t) = u(x,t_0) \, e^{i\omega t} \tag{2.50}$$

where $u(x,t_0)$ is the amplitude of the function at the initial time t_0. It should be noted that u(x,t) is exactly defined at any time, t, provided its initial amplitude is known. It constitutes, therefore, a ground truth for comparison with solutions of (2.47) obtained using any time differencing scheme. Furthermore, since the amplitude of the wave is bounded by $|u(x,t_0)|$, any variation of u(x,t) outside these limits during the integration of (2.49) using a numerical method is attributed to the integration scheme. It is therefore important to design time differencing schemes that do not amplify the solution. In finite form, (2.50) may be rewritten to give

$$u(x,n\Delta t) = u(x,t_0) \, e^{i\omega \, n\Delta t} \tag{2.51}$$

where n is the time level. If the spatial coordinate is omitted, as only time integration is considered, (2.51) may finally be expressed as

$$u(n\Delta t) = u(t_0) \, e^{i\omega \, n\Delta t} \tag{2.52}$$

In order to determine the stability of the scheme, an amplification factor,

λ, is introduced such as

$$u^{n+1} = \lambda u^n \tag{2.53}$$

The stability of the scheme is then defined as

$$
\begin{array}{lll}
\text{stable} & |\lambda| < 1 & \\
\text{neutral} & |\lambda| = 0 & \tag{2.54} \\
\text{unstable} & |\lambda| > 1 &
\end{array}
$$

Using this simple, but important, linear differential equation, the stability of several classical time differencing schemes is discussed in the following sections. It should be noted that the formulation of the time differencing scheme is of central importance in the coding of a time dependent model.

6.1 Euler, the Backward and the Trapezoidal Schemes

The basic concept of time integration is to predict the value of a time dependent function at time level (n+1) given its value at time level n. Therefore, rewriting (2.49) as

$$\frac{du(t)}{dt} = F(u,t) \tag{2.55}$$

and integrating between time $n\Delta t$ and $(n+1)\Delta t$ one obtains

$$u^{n+1} = u^n + \int_{n\Delta t}^{(n+1)\Delta t} F(u,t)\, dt \tag{2.56}$$

where $F(u,t)$ is the forcing function and takes the values of

$$
\begin{cases}
F = F^n & \text{at } t = n\Delta t \\[2mm]
F = F^{n+1} & \text{at } t = (n+1)\Delta t
\end{cases} \tag{2.57}
$$

If F^{n+1} is a function of u^{n+1} the scheme is called implicit; otherwise, it is explicit. Within the interval $(n\Delta t, (n+1)\Delta t)$, $F(u,t)$ can also be represented by a combination of its values at time levels n and $(n+1)$ as

$$F = \alpha F^n + \beta F^{n+1} \tag{2.58}$$

In this case, (2.56) may be written as

$$u^{n+1} = u^n + \Delta t \left[\alpha F^n + \beta F^{n+1} \right] \qquad \text{with } \alpha + \beta = 1 \tag{2.59}$$

where different values of α and β lead to different schemes. For example,

$\alpha = 1, \beta = 0$ Euler (forward) scheme
$\alpha = 0, \beta = 1$ Backward scheme
$\alpha = 1/2, \beta = 1/2$ Trapezoidal scheme

Substituting for F in (2.59) yields

$$u^{n+1} = u^n + \Delta t \left[\alpha(i\omega u^n) + \beta(i\omega u^{n+1}) \right] \tag{2.60}$$

or

$$u^{n+1} = \frac{1 + i\alpha\omega\Delta t}{1 - i\beta\omega\Delta t} u^n \tag{2.61}$$

The amplification factor is then obtained as

$$|\lambda| = \left[\frac{(1-\alpha\beta p^2)^2 + (\alpha+\beta)^2 p^2}{(1+\beta^2 p^2)^2} \right]^{\frac{1}{2}} \tag{2.62}$$

where $p = \omega\Delta t$.

6.1.1 Euler Scheme

In this case $\alpha = 1$, $\beta = 0$, and the amplification factor is given by

$$|\lambda| = \left[1 + p^2 \right]^{\frac{1}{2}} \tag{2.63}$$

Since p^2 is always positive, λ is always greater than unity and the scheme is said to be unstable.

6.1.2 Backward Scheme

Here $\alpha = 0$ and $\beta = 1$, and the amplification factor reduces to

$$|\lambda| = \left[1 + p^2\right]^{-\frac{1}{2}}$$

(2.64)

This scheme is unconditionally stable as $|\lambda|$ is always less than unity. Furthermore, the amplification factor decreases as the wave frequency increases, producing more damping for the high frequency modes. The use of the Backward scheme is then desirable at the start of models integration to reduce the amplitude of the gravity waves.

6.1.3 Trapezoidal Scheme

For this scheme $\alpha = \beta = 1/2$ and the amplification factor is given by,

$$|\lambda| = 1$$

(2.65)

Such schemes are called neutral or non-amplifying schemes.

6.2 Matsuno and Heun's Schemes

These type of schemes are called predictor-corrector schemes and are used in a two step method. Here also the basic wave equation is used to implement the schemes and to discuss their stability. That is,

$$\frac{du}{dt} = F$$

(2.66)

This equation is integrated in two sequences. First the prediction of u^{n+1} is done and is designated by $u^{(n+1)^*}$. The second step consists of evaluating F^{n+1} using $u^{(n+1)^*}$. Finally $F^{(n+1)^*}$ is used to improve the initial value of u^{n+1}. The finite difference formulation of this procedure is written as, predictor step $u^{(n+1)^*} = u^n + \Delta t\, F^n$ (2.67)

corrector step $u^{n+1} = u^n + \Delta t\left[\alpha\, F^n + \beta\, F^{(n+1)^*}\right]$ (2.68)

where, $F^{(n+1)^*}$ is obtained using $u^{(n+1)^*}$.

Substituting for F and combining (2.67) with (2.68) one gets,

$$u^{n+1} = u^n + i\omega\Delta t\,\{\alpha u^n + \beta[u^n + i\omega\Delta t u^n]\}$$ (2.69)

or

$$u^{n+1} = \{[1 - \beta\omega^2\Delta t^2] + i\,[(\alpha + \beta)\,\omega\Delta t]\}\,u^n$$ (2.70)

Since $(\alpha + \beta) = 1$, the amplification factor for these types of schemes may then be expressed as

$$|\lambda| = \left[(1 - \beta\omega^2\Delta t^2)^2 + \omega^2\Delta t^2\right]^{\frac{1}{2}}$$ (2.71)

6.2.1 Matsuno Scheme

For this scheme $\alpha = 0$ and $\beta = 1$. Thus the amplifying factor reduces to

$$|\lambda| = \left\{(1 - p^2)^2 + p^2\right\}^{\frac{1}{2}}$$ (2.72)

or

$$|\lambda| = p^4 - p^2 + 1$$ (2.73)

and the stability condition is expressed as

$$|\lambda| \le 1 \text{ if } \omega\Delta t \le 1.$$ (2.74)

This scheme is not suitable for high frequency modes.

6.2.2 Heun's Scheme

In this case $\alpha = \frac{1}{2}$ and $\beta = \frac{1}{2}$, and the stability condition is given by

$$|\lambda| = \left[1 + \frac{p^4}{4}\right]^{\frac{1}{2}} \tag{2.75}$$

Heun's scheme is then unconditionally unstable.

6.3 Adams Bashforth Scheme

This is a three time level scheme whose difference formulation is given by

$$u^{n+1} = u^n - ikc\Delta t \left[\frac{3}{2} u^n - \frac{1}{2} u^{n-1}\right] \tag{2.76}$$

In this formulation, the space derivative of the function is obtained as a linear combination of its values at time levels n and n−1. Substituting for u^{n+1} and u^n the following form is obtained for the stability condition,

$$\lambda^2 - \lambda (1 + i\frac{3}{2} \omega\Delta t) + \frac{1}{2} i\omega\Delta t = 0 \tag{2.77}$$

where $\omega = -kc$. This is a quadratic equation in λ and has two roots,

$$\lambda_1 = \frac{1}{2}\left[(1 + i\frac{3}{2} \omega\Delta t) + \sqrt{1 - \frac{9}{4} \omega^2\Delta t^2 + i\omega\Delta t}\right] \tag{2.78}$$

and

$$\lambda_2 = \frac{1}{2}\left[(1 + i\frac{3}{2} \omega\Delta t) - \sqrt{1 - \frac{9}{4} \omega^2\Delta t^2 + i\omega\Delta t}\right] \tag{2.79}$$

It is clear that this scheme is unstable for large time steps. However, as Δt decreases and approaches zero, $\lambda_1 \to 1$ and $\lambda_2 \to 0$. λ_1 is the amplification factor for the physical solution whereas λ_2 represents that of the computational mode.

6.4 Leap Frog Scheme

This is also a three time level scheme which uses the present and past values of the function to predict its future state. Furthermore, the Leap Frog is centered in space, and is therefore also called centered in space, centered in time scheme. It is one of the most popular schemes in numerical weather prediction.

The solution of the linear equation may then be discretized as

$$u(m\Delta x, n\Delta t) = u(n\Delta t)\, e^{ikm\Delta x} \tag{2.80}$$

where $m\Delta x = x$ and $n\Delta t = t$. Substitution into the wave equation using centered differences in space and time yields

$$\frac{u_m^{n+1} - u_m^{n-1}}{2\Delta t} = -c\, \frac{u_{m+1}^n - u_{m-1}^n}{2\Delta x} \tag{2.81}$$

or

$$u_m^{n+1} = u_m^{n-1} - \frac{c\Delta t}{\Delta x}[u_{m+1}^n - u_{m-1}^n] \tag{2.82}$$

Upon substitution for u one obtains

$$u^{n+1} e^{ikm\Delta x} - u^{n-1} e^{ikm\Delta x} = -\frac{c\Delta t}{\Delta x}[u^n e^{ik(m+1)\Delta x} - u^n e^{ik(m-1)\Delta x}] \tag{2.83}$$

which can be reduced to

$$u^{n+1} - u^{n-1} = -\frac{2ic\Delta t}{\Delta x} \sin(k\Delta x)\, u^n \tag{2.84}$$

Using (2.53) the stability equation is formed as

$$\lambda^2 + \lambda\, \frac{2ic\Delta t}{\Delta x} \sin(k\Delta x) - 1 = 0 \tag{2.85}$$

whose solutions are

$$\lambda_1 = \left[1 - \frac{c^2\Delta t^2}{\Delta x^2} \sin^2(k\Delta x) \right]^{\frac{1}{2}} - i\, \frac{c\Delta t}{\Delta x} \sin(k\Delta x) \tag{2.86}$$

and

$$\lambda_2 = -\left[1 - \frac{c^2\Delta t^2}{\Delta x^2} \sin^2(k\Delta x) \right]^{\frac{1}{2}} - i\, \frac{c\Delta t}{\Delta x} \sin(k\Delta x) \tag{2.87}$$

Thus,

$$|\lambda_1| = |\lambda_2| = 1 \tag{2.88}$$

and the scheme is neutral. It is important to note, however, that as Δt tends to zero, λ_1 tends to 1 representing the physical solution and λ_2 tends to -1 representing the computational solution. Furthermore, it is easy to see that for the quantity under the radical to be positive, the following condition must be satisfied

$$\frac{c\Delta t}{\Delta x} \leq 1 \tag{2.89}$$

This is known as the Courant-Friedrichs-Levy (CFL) condition.

6.5 Implicit Schemes

Implicit schemes are more economic than their explicit counterpart in the sense that they allow for time steps much larger than those required by the CFL condition. They are also known to damp the amplitude of the fast moving gravity waves. In these schemes, the space derivative at time level n is evaluated as a mean value between the space derivatives at times $(n+1)$ and $(n-1)$. This technique is equivalent to evaluating the time derivative at half time step. Since the scheme assumes implicitly an unknown future value, it is called implicit.

6.5.1 Fully implicit scheme

In implicit form, the finite difference analog of the linear equation is written as

$$u_m^{n+1} - u_m^n = -\frac{c\Delta t}{2}\left[\frac{u_{m+1}^{n+1} - u_{m-1}^{n+1}}{2\Delta x} + \frac{u_{m+1}^n - u_{m-1}^n}{2\Delta x}\right] \tag{2.90}$$

It is to be noted that all variables at time level $(n+1)$ are unknown, and these unknowns involve three different locations. In principle, for linear problems, the system can be solved by matrix inversion provided appropriate boundary conditions. This method is not desirable for a large number of grid point and the relaxation method is then preferred.

If the solution $u_m^n = u^n\, e^{ikm\Delta x}$ is adopted, (2.90) takes the form,

$$u_m^{n+1} - u_m^n = -i\,\frac{c\Delta t}{2}\sin k\Delta x(u_m^{n+1} - u_m^n) \tag{2.91}$$

Since $u_m^{n+1} = \lambda u_m^n$, then one obtains

$$\lambda = \frac{1 - i \, \dfrac{c\Delta t}{\Delta x} \, \sin k\Delta x}{1 + i \, \dfrac{c\Delta t}{\Delta x} \, \sin k\Delta x} \tag{2.92}$$

and $|\lambda| = 1$ regardless of the value of $\dfrac{c\Delta t}{\Delta x}$. This scheme is uncon-
ditionally stable.

6.5.2 Semi-implicit Scheme

In the semi-implicit time integration formulation, the fast and slow
moving waves are separated. The low frequency modes are treated
explicitly while the high frequency modes are handled implicitly. Let

$$\omega = \alpha + \beta \quad \text{and} \quad \beta > \alpha \tag{2.93}$$

where α and β represent the low and high frequency parts of the wave,
respectively. The finite difference form of $\dfrac{\partial u}{\partial t} = i\omega u$ is therefore expressed as

$$\frac{u_m^{n+1} - u_m^{n-1}}{2\Delta t} = i\alpha u_m^n + i\beta \, \frac{u_m^{n+1} + u_m^{n-1}}{2} \tag{2.94}$$

where the first and second terms of the right hand side represent the
explicit and implicit parts, respectively.

Equation (2.94) can be rewritten as

$$u_m^{n+1} = u_m^{n-1} + 2\Delta t \left[i\alpha u_m^n + i\beta \, \frac{u_m^{n+1} + u_m^{n-1}}{2} \right] \tag{2.95}$$

Note that if $\alpha = 0$, the unknown u_m^{n+1} appears in both sides of the
equation and the scheme would be fully implicit,

$$u_m^{n+1} = u_m^{n-1} + i\beta\, 2\Delta t \left[\frac{u_m^{n+1} + u_m^{n-1}}{2} \right] \qquad (2.96)$$

On the other hand if $\beta = 0$, the unknown appears only on the left hand side of the equation suggesting a fully explicit scheme,

$$u_m^{n+1} = u_m^{n-1} + i2\alpha\Delta t u_m^{n} \qquad (2.97)$$

Implicit schemes are always stable and allow for large time steps. Semi-implicit schemes also allow larger time steps than most explicit schemes.

If the solution $u(m\Delta x, n\Delta t) = u^n e^{ikm\Delta x}$ is assumed, (2.95) takes the form

$$u^{n+1} = u^{n-1} + 2\Delta t \left[i\alpha u^{n} + i\beta \left[\frac{u^{n+1} + u^{n-1}}{2} \right] \right] \qquad (2.98)$$

using $u^{n+1} = \lambda u^{n}$, a quadratic equation for λ can be formed as

$$(1 - i\beta\Delta t)\lambda^2 - 2i\alpha\Delta t\lambda - (1 + i\beta\Delta t) = 0 \qquad (2.99)$$

The roots of this equation are

$$\lambda_1 = \frac{2i\alpha\Delta t + \sqrt{4(1+\beta^2\Delta t^2) - 4\alpha^2\Delta t^2}}{2(1 - i\beta\Delta t)} \qquad (2.100)$$

and

$$\lambda_2 = \frac{2i\alpha\Delta t - \sqrt{4(1+\beta^2\Delta t^2) - 4\alpha^2\Delta t^2}}{2(1 - i\beta\Delta t)} \qquad (2.101)$$

Note that, if $\Delta t \to 0$, $\lambda_1 \to 1$ and $\lambda_2 \to -1$. Thus λ_1, denotes the physical mode, and λ_2 is the computational solution. In fact λ_1 can be rewritten as

$$\lambda_1 = \frac{(2i\omega_2 + 2\sqrt{1 + \omega_1^2 - \omega_2^2})(1 + i\omega_1)}{2(1+\omega_1^2)} \tag{2.102}$$

where $\omega_1 = \beta\Delta t$ and $\omega_2 = \alpha\Delta t$.

Finally, simple manipulation of (2.102) shows that the amplification factor, $|\lambda| = 1$ and that the scheme is stable.

3

Calculation of Vertical Motion

The vertical wind velocity is not an observed variable in meteorology, and its estimation appears as one of the most difficult problems. The vertical velocity is an integral component of the three dimensional structure of the atmospheric motion and is encountered in many diagnostic and prognostic problems. The simplest method for the computation of the vertical motion would be the integration of the mass continuity equation using the large scale horizontal wind observations and accounting for the divergence correction. The sparseness of observations constitutes, however, a serious obstacle in this so-called kinematic method. Moreover, due to uncertainties inherent to wind measurements, large errors are introduced in the calculation of the horizontal divergence and lead to important errors in the estimation of the vertical velocity.

Besides the kinematic vertical velocity, several other methods have been developed to compute the vertical motion of the atmosphere. Among these, one can mention the adiabatic method which is based on the thermodynamic energy equation and is not very sensitive to errors in the measured wind field. In this case the temperature advection can be estimated quite accurately using the geostrophic wind, especially in mid-latitudes where observations are dense. This method can then be employed when only geopotential and temperature data are available. Nevertheless, the adiabatic method involves the temperature tendency and is not recommended for a wide area unless observations are not too far apart in time. The vertical velocity can also be estimated using a simplified form of the vorticity equation. In this so-called vorticity method, the vertical advection of vorticity and the "twisting" term are neglected and the relative vorticity is assumed to be small as compared to the Coriolis parameter in the divergence term. The time tendency and the horizontal advection of the geostrophic vorticity can then be estimated with a reasonable degree of accuracy. This leads to vertical velocity estimates more reliable than those obtained using the kinematic method. Finally, the vertical velocity can also be obtained using the quasi-geostrophic omega equation. This method is strictly diagnostic and estimates the vertical motion in terms of instantaneous values of geopotential. Furthermore, the use of the quasi-geostrophic omega equation does not require wind observations nor does it involve time tendencies, which makes it superior to all other methods.

In this chapter, different techniques for estimating the vertical motion are developed. Both the cases where data are regularly and irregularly spaced over a horizontal domain are considered.

1. Vertical Velocity from Irregularly Spaced Wind Data

One of the techniques used to estimate the vertical velocity from irregularly spaced wind observations is the polynomial method. This method has been described by Yanai et al. (1973), and consists of fitting a polynomial to data from an irregularly distributed observational network. The technique is based on the least squares approximation.

1.1 Triangular Representation

This section illustrates an example of the estimation of the vertical velocity using the triangular method over a network having three weather stations. In this case, the zonal and meridional components of the wind are expressed as linear functions of the observing site locations.

$$u = ax + by + c \tag{3.1}$$
$$v = px + qy + r \tag{3.2}$$

The technique consists of determining the coefficients a, b, c, p, q, and r, using the least squares approach. This is done through the minimization of the error sum and the solution of the following normal equations,

$$cN + a\sum x_i + b\sum y_i = \sum u_i \tag{3.3}$$
$$c\sum x_i + a\sum x_i^2 + b\sum x_i y_i = \sum x_i u_i \tag{3.4}$$
$$c\sum y_i + a\sum x_i y_i + b\sum y_i^2 = \sum y_i u_i \tag{3.5}$$

The variables x_i and y_i define the locations of the stations. The solution of this system gives a, b, and c. The coefficients p, q, and r for the meridional component are obtained following the same procedure. In this formulation the divergence and vorticity are defined in a simple manner. For example, the divergence is expressed as

$$\left[\frac{\partial u}{\partial x} + \frac{\partial v}{\partial y} \right] = a + q \tag{3.6}$$

and the vorticity is given by

$$\left[\frac{\partial v}{\partial x} - \frac{\partial u}{\partial y} \right] = p - b \tag{3.7}$$

The coefficients c and r represent the translation of the wind with respect to the origin of the network (x = y = 0). In a multilevel problem, the computation of the least squares coefficients must be performed at each pressure level. Because of the inherent errors, the vertical integral of the divergence is generally large and does not satisfy the Dynes compensation. In order to meet this condition, the error in the divergence is assumed to be proportional to its magnitude as

$$\left. \vec{\nabla} \cdot \vec{V} \right]_c = \left. \vec{\nabla} \cdot \vec{V} \right]_u + \varepsilon \left| \left. \vec{\nabla} \cdot \vec{V} \right| \right._u = - \frac{\partial \omega}{\partial p} \tag{3.8}$$

where u stands for uncorrected and c for corrected divergence. The term ω represents the vertical velocity in the pressure coordinate system and is expressed in mbs^{-1}; p is the atmospheric pressure.

The mass continuity requirement is imposed on the corrected divergence to provide the correction factor,

$$\int_{1000}^{0} \left. \vec{\nabla} \cdot \vec{V} \right]_c dp = 0 \tag{3.9}$$

and

$$\varepsilon = - \frac{\displaystyle\int_{1000}^{0} \left. \vec{\nabla} \cdot \vec{V} \right]_u dp}{\displaystyle\int_{1000}^{0} \left| \vec{\nabla} \cdot \vec{V} \right|_u dp} \tag{3.10}$$

Assuming the vertical velocity to be zero at the lowest level, at any other pressure level, it is obtained by integrating (3.8),

$$\omega\,(p) = -\int_0^p \nabla.\vec{V}\;dp - \frac{\displaystyle\int_{1000}^{0} \nabla.\vec{V}\;dp}{\displaystyle\int_{1000}^{0} |\nabla.\vec{V}|\;dp}\int_0^p |\nabla.\vec{V}|\;dp \qquad (3.11)$$

A computer code illustrating an example of calculation of the kinematic vertical velocity by the linear multiple regression approach is provided (**KINEMATIC**). Solutions of the linear systems are computed by subroutine **PLNSFC**. The output from this method is shown in Table 3.1.

Table 3.1: Computation of the vertical velocity using the kinematic method.

Pressure level (mb)	Omega (mb/sec)	Divergence (per sec)	Vorticity (per sec)
1000	0.0000E+00		
		0.6218E−04	0.2963E−03
900	0.6218E−02		
		−0.1729E−03	0.1551E−03
800	−0.1107E−01		
		−0.1286E−03	−0.2089E−03
700	−0.2392E−01		
		−0.1267E−03	−0.3603E−03
600	−0.3659E−01		
		0.1012E−03	−0.3871E−03
500	−0.2638E−01		
		0.7086E−04	−0.4375E−03
400	−0.1930E−01		
		0.1930E−03	−0.5423E−03
300	0.1863E−08		

program *KINEMATIC*

```
c
c       this program computes the kinematic vertical motion using a tria
c       -ngular method. the observations need to be stratified in the ve
c       -rtical each layer must have at least three observations not on a
c       straight line. the layers are seperated by a check of alat = 999.
c       the subroutine has 10 levels and a maximum of 75 point values of
c       direction and speed measured in degrees and knots, respectively.
c       these observations are irregularly spaced.
c       the vertical motion, w is computed for the 10 levels using the
c       mean divergence of the layer.omega at the surface is set to zero
c       whereas at the top it is computed assuming that the layer diverg
c       -ence is the same as the layer below it.
c
c               definitions of parameters
c
c       nlvl        : number of stratified layers
c       nrec        : number of obs in the vertical layer (max 75)
c       pres        : pressure level in mb (f6.1)
c       idir        : wind direction in degrees (i3)
c       ispd        : wind speed (knots,i3)
c       xo          : longitude of the origin in degrees
c       yo          : latitude of the origin in degrees
c       alat        : latitudinal position of obs (f5.1)
c       alon        : longitudinal position of obs (f5.1)
c       au,bu,cu    : coefficients of planar surface of x-component
c       av,bv,cv    : coefficients of planar surface of y-component
c       div         : mean layer divergence (per sec)
c       vort        : mean layer vorticity (per sec)
c       w           : vertical velocity (mb/sec)
c       fact        : conversion factor from knots to meter per sec
c       rpd         : conversion factor from degrees to radians
c       npt         : number of observations in a layer
c       divs        : vertical integral of the mean divergence
c       diva        : magnitude of the mean divergence
c       eps         : correction factor of the mean divergence
c       dp          : depth of the pressure levels (here constant)
c
        parameter (nlvl = 8 ,nrec = 75)
c
c       declaration of varibles
c
        real div(8), vort(8), w(10), p1(10), ucomp(75) , vcomp(75)
```

```
         real ylat(75),xlong(75),au(8),bu(8),cu(8),av(8),bv(8),cv(8)
c
         open(5,file='kinematic.dat',status='old')
c
         data fact,xo,yo,dp/0.514791,42.,15.,100./
         pi          = 4.0*atan(1.0)
         divj        = 0.
         divi        = 0.
         rpd         = pi/180.
c
c     initialize the vertical motion to zero
c
         do 3100 nk = 1, 10
 3100    w(nk)       = 0.0
c
c     loop over levels
c
         do 3102 nk = 1, nlvl
             npt     = 0
         do 3104 n  = 1, nrec
c
c     read in observations with specified format
c
         read(5,15,end=22)alat,alon,pres,idir,ispd
    24   continue
         if (alat.eq.999.) go to 13
         spd         = float(ispd)*fact
         beta        = float(idir)*rpd
         ucomp(n)    = -spd*sin(beta)
         vcomp(n)    = -spd*cos(beta)
         ylat (n)    = (alat-yo)*1.111e5
         xlong(n)    = (alon-xo)*1.111e5*cos(alat)*rpd
         npt         = npt+1
 3104    continue
    13   continue
c
c     subroutine PLNSFC is called for both u- and v-component
c     to solve the linear system in each layer.
c
         call PLNSFC (ucomp,xlong,ylat,au(nk),bu(nk),cu(nk),npt)
         call PLNSFC (vcomp,xlong,ylat,av(nk),bv(nk),cv(nk),npt)
 3102    continue
    22   continue
```

```
c
c      compute divergence and vorticity in the layer. Vorticity is added
c      just for complement. Equations (3.6) and (3.7) are used.
c
           do 3106 k = 1, nlvl
               vort(k) = av(k)-bu(k)
               div(k)  = au(k)+bv(k)
               divs    = divs+div(k)
               diva    = diva+abs(div(k))
     3106  continue
c
c      compute the divergence in the ninth layer.
c
           divs       = divs+div(nlvl)
           diva       = diva+abs(div(nlvl))
c
c      correct the divergence of the layer.
c
           eps        = -(divs/diva)
           do 3108 k = 1, nlvl
               div(k)  = div(k)+eps*abs(div(k))
     3108  continue
c
c      compute the omegas at all levels
c
           w(1)       = 0.
           do 3110 k = 2, 9
     3110  w(k)       = w(k-1)+div(k-1)*dp
           w(nlvl+2)  = w(nlvl+1)+div(nlvl)*dp
c
c      write output for the 8 first levels.
c      the format shows that the divergence
c      and vorticity are computed in the layers
c      between two pressure levels.
c
           do 3112 k = 1, 10
     3112  pl(k) = 1000.-((k-1)*100.)
           write(6,1000)
           write(6,1001)
           do 3114 k = 1, nlvl
           write (6,1002) pl(k) , w(k)
           if(k.ge.8) goto 3114
           write(6,1003) div(k), vort(k)
```

```
3114    continue
  15    format(2f5.1,f6.1,1x,2i3)
1000    format(//,1x,'pres lvl',4x,' omega ',6x,'divergence',
   +    5x,'vorticity')
1001    format(3x,'(mb)',5x,'(mb/sec)',6x,'(per sec)',
   +    6x,'(per sec)',/)
1002    format(1x,/,2x,f5.0,2x,e12.4)
1003    format(24x,e12.4,2x,e12.4)
        stop
        end
```

1.2 Quadratic Representation

The quadratic representation is often used in geophysical sciences to fit a surface to a given set of observations. In this method, the functional approximation has the following form,

$$f(x,y) = a_0 + a_1 x + a_2 y + a_3 xy + a_4 x^2 + a_5 y^2 \qquad (3.12)$$

This function is exactly defined with a set of six observations. The solution to this problem consists of determining the coefficients $\{a_j\}$. The standard procedure is to minimize the discrepancy sum,

$$E = \sum_{i=1}^{N} [f(x_i,y_i) - \tilde{f}(x_i,y_i)]^2 \qquad (3.13)$$

or,

$$E = \sum_{i=1}^{N} [a_0 + a_1 x_i + a_2 y_i + a_3 x_i y_i + a_4 x_i^2 + a_5 y_i^2 - \tilde{f}(x_i,y_i)]^2 \qquad (3.14)$$

The set of variables $\{x_i\}$ and $\{y_i\}$ represents the station locations, and the functions $\tilde{f}(x_i,y_i)$ are the observed field. The coefficients are obtained as solutions to the normal equations generated by the minimization of the error sum of squares,

$$\frac{\partial E}{\partial a_j} = 0 \qquad (j = 0, \dots , 5) \qquad (3.15)$$

This leads to

$$\frac{\partial E}{\partial a_0} = 2 \sum_{i=1}^{N} [a_0 + a_1 x_i + a_2 y_i + a_3 x_i y_i + a_4 x_i^2 + a_5 y_i^2 - \tilde{f}(x_i, y_i)] = 0$$

$$(3.16)$$

$$\frac{\partial E}{\partial a_1} = 2 \sum_{i=1}^{N} [a_0 + a_1 x_i + a_2 y_i + a_3 x_i y_i + a_4 x_i^2 + a_5 y_i^2 - \tilde{f}(x_i, y_i)] x_i = 0$$

$$(3.17)$$

$$\frac{\partial E}{\partial a_2} = 2 \sum_{i=1}^{N} [a_0 + a_1 x_i + a_2 y_i + a_3 x_i y_i + a_4 x_i^2 + a_5 y_i^2 - \tilde{f}(x_i, y_i)] y_i = 0$$

$$(3.18)$$

$$\frac{\partial E}{\partial a_3} = 2 \sum_{i=1}^{N} [a_0 + a_1 x_i + a_2 y_i + a_3 x_i y_i + a_4 x_i^2 + a_5 y_i^2 - \tilde{f}(x_i, y_i)] x_i y_i = 0$$

$$(3.19)$$

$$\frac{\partial E}{\partial a_4} = 2 \sum_{i=1}^{N} [a_0 + a_1 x_i + a_2 y_i + a_3 x_i y_i + a_4 x_i^2 + a_5 y_i^2 - \tilde{f}(x_i, y_i)] x_i^2 = 0$$

$$(3.20)$$

$$\frac{\partial E}{\partial a_5} = 2 \sum_{i=1}^{N} [a_0 + a_1 x_i + a_2 y_i + a_3 x_i y_i + a_4 x_i^2 + a_5 y_i^2 - \tilde{f}(x_i, y_i)] y_i^2 = 0$$

$$(3.21)$$

Because of the errors inherent to observations and their variations from one observation to another, it is often preferable to minimize a weighted sum of squares. The weights are intended to reflect accuracy variations in the data set and are generally taken as

$$\alpha_i^2 = \frac{1}{2\sigma_i^2} \qquad (3.22)$$

where σ_i^2 is the variance of the set of observations. Thus, if the observation is noisy (σ_i^2 large), it receives a small weight. The discrepancy sum is then rewritten as

$$E = \sum_{i=1}^{N} \alpha_i^2 [f(x_i, y_i) - \tilde{f}(x_i, y_i)]^2 \qquad (3.23)$$

Introducing the operator

$$\overline{(q)} = \sum_{i=1}^{N} (q) \tag{3.24}$$

where q can be any scalar quantity and using a matrix notation, (3.23) reduces to

$$AX = b \tag{3.25}$$

where A is the coefficients matrix and is given by

$$
A = \begin{bmatrix}
\overline{\alpha_i^2} & \overline{\alpha_i^2 x_i} & \overline{\alpha_i^2 y_i} & \overline{\alpha_i^2 x_i y_i} & \overline{\alpha_i^2 x_i^2} & \overline{\alpha_i^2 y_i^2} \\
\overline{\alpha_i^2 x_i} & \overline{\alpha_i^2 x_i^2} & \overline{\alpha_i^2 x_i y_i} & \overline{\alpha_i^2 x_i^2 y_i} & \overline{\alpha_i^2 x_i^3} & \overline{\alpha_i^2 x_i y_i^2} \\
\overline{\alpha_i^2 y_i} & \overline{\alpha_i^2 x_i y_i} & \overline{\alpha_i^2 y_i^2} & \overline{\alpha_i^2 x_i y_i^2} & \overline{\alpha_i^2 x_i^2 y_i} & \overline{\alpha_i^2 y_i^3} \\
\overline{\alpha_i^2 x_i y_i} & \overline{\alpha_i^2 x_i^2 y_i} & \overline{\alpha_i^2 x_i y_i^2} & \overline{\alpha_i^2 x_i^2 y_i^2} & \overline{\alpha_i^2 x_i^3 y_i} & \overline{\alpha_i^2 x_i y_i^3} \\
\overline{\alpha_i^2 x_i^2} & \overline{\alpha_i^2 x_i^3} & \overline{\alpha_i^2 x_i^2 y_i} & \overline{\alpha_i^2 x_i^3 y_i} & \overline{\alpha_i^2 x_i^4} & \overline{\alpha_i^2 x_i^2 y_i^2} \\
\overline{\alpha_i^2 y_i^2} & \overline{\alpha_i^2 x_i y_i^2} & \overline{\alpha_i^2 y_i^3} & \overline{\alpha_i^2 x_i y_i^3} & \overline{\alpha_i^2 x_i^2 y_i^2} & \overline{\alpha_i^2 y_i^4}
\end{bmatrix} \tag{3.26}
$$

X is the unknown vector defined as

$$
X = \begin{bmatrix}
\overline{a_0} \\
\overline{a_1} \\
\overline{a_2} \\
\overline{a_3} \\
\overline{a_4} \\
\overline{a_5}
\end{bmatrix} \tag{3.27}
$$

and b is the forcing vector given by

$$b = \begin{bmatrix} \overline{\alpha_i^2 \; \tilde{f}(x_i,y_i)} \\ \overline{\alpha_i^2 \; \tilde{f}(x_i,y_i) \; x_i} \\ \overline{\alpha_i^2 \; \tilde{f}(x_i,y_i) \; y_i} \\ \overline{\alpha_i^2 \; \tilde{f}(x_i,y_i) \; x_i y_i} \\ \overline{\alpha_i^2 \; \tilde{f}(x_i,y_i) \; x_i^2} \\ \overline{\alpha_i^2 \; \tilde{f}(x_i,y_i) \; y_i^2} \end{bmatrix} \qquad (3.28)$$

This system can be solved by a matrix inversion algorithm. It should be noted that the coefficient matrix is symmetric, and this can reduce the computational burden of solution. But it is not well conditioned. Therefore, unless special fitting conditions are introduced, least-squares fitting beyond the sixth or seventh degree polynomial results in computational difficulties. At the issue of this procedure, a two dimensional quadratic surface can be fitted to any scalar field.

This algorithm can be extended and used for the determination of the vertical velocity in a pressure coordinate system. The vertical integral of the continuity equation gives

$$\omega \, (p) = \omega \, (p_0) - \int\limits_{P_0}^{p} \left[\frac{\partial u}{\partial x} + \frac{\partial v}{\partial y} \right] dp \qquad (3.29)$$

Using the quadratic representation, the zonal and meridional components of the wind estimates are expressed as

$$\tilde{u}(x,y) = a_0 + a_1 x + a_2 y + a_3 xy + a_4 x^2 + a_5 y^2 \qquad (3.30)$$

$$\tilde{v}(x,y) = b_0 + b_1 x + b_2 y + b_3 xy + b_4 x^2 + b_5 y^2 \qquad (3.31)$$

The estimated divergence is then given by

$$\frac{\partial \tilde{u}}{\partial x} + \frac{\partial \tilde{v}}{\partial y} = (a_1 + b_2) + (b_3 + 2a_4)x + (a_3 + 2b_5)y \qquad (3.32)$$

Assuming a least-squares analysis of the horizontal wind has been carried out at the appropriate pressure levels, the vertical velocity is estimated at any pressure level using

$$\omega(p) = \omega(p_0) - \int_{p_0}^{p} \left\{ (a_1 + b_2) + (b_3 + 2a_4)x + (a_3 + 2b_5)y \right\} dp \qquad (3.33)$$

The value of the vertical velocity at the lower boundary of the atmosphere can be obtained by performing a least-squares analysis of the terrain field, h, and then requiring

$$\omega(p_0) = - \rho_0 \, g \, \overrightarrow{V_0} . \nabla h \qquad (3.34)$$

where ρ_0 and $\overrightarrow{V_0}$ are, respectively, the air density and the wind speed at the surface. In case a flat surface is assumed, $\omega(p_0)$ is set to zero. Due to errors in the wind observations, the mass continuity equation is generally not satisfied. That is,

$$\int_{p_0}^{0} \left[\frac{\partial \tilde{u}}{\partial x} + \frac{\partial \tilde{v}}{\partial y} \right] dp \neq 0 \qquad (3.35)$$

In terms of least-squares coefficients, this condition takes the form,

$$\int_{p_0}^{0} \left\{ (a_1 + b_2) + (b_3 + 2a_4)x + (a_3 + 2b_5)y \right\} dp \neq 0 \qquad (3.36)$$

If the mass continuity condition were to be satisfied, a new set of coefficients need to be determined such that

$$\int_{p_0}^{0} \left\{ (a_1^* + b_2^*) + (b_3^* + 2a_4^*)x + (a_3^* + 2b_5^*)y \right\} dp = 0 \qquad (3.37)$$

It should be noted that (3.37) may be satisfied only at particular locations (x,y) and not necessarily over the entire domain. Alternately (3.37) can be integrated over a large domain, but in this case, the mass balance is satisfied over the averaged domain but not necessarily at any individual point. In this last formulation, a consistent set of analyzed values of horizontal wind, divergence and vertical velocity will be carried out at the center of gravity (x_c, y_c) of a finite network of observation. Assuming the error in the divergence to be proportional to its magnitude, one can write

$$(a_1^* + b_2^*) + (b_3^* + 2a_4^*)x_c + (a_3^* + 2b_5^*)y_c = (a_1 + b_2) + (b_3 + 2a_4)x_c$$
$$+ (a_3 + 2b_5)y_c + \varepsilon \left| (a_1 + b_2) + (b_3 + 2a_4)x_c + (a_3 + 2b_5)y_c \right| \quad (3.38)$$

Integrating (3.38) vertically yields

$$\varepsilon = \cfrac{-\int\limits_{p_0}^{0} \left\{ (a_1 + b_2) + (b_3 + 2a_4)x_c + (a_3 + 2b_5)y_c \right\} dp}{\int\limits_{p_0}^{0} \left| (a_1 + b_1) + (b_3 + 2a_4)x_c + (a_3 + 2b_5)y_c \right| dp} \quad (3.39)$$

This error is assumed to be equally distributed among all the terms. The final set of coefficients is then given by

$$a_1^* = a_1 + \frac{\varepsilon}{6} \left| (a_1 + b_2) + (b_3 + 2a_4)x_c + (a_3 + 2b_5)y_c \right| \quad (3.40)$$

$$b_2^* = b_2 + \frac{\varepsilon}{6} \left| (a_1 + b_2) + (b_3 + 2a_4)x_c + (a_3 + 2b_5)y_c \right| \quad (3.41)$$

$$a_3^* = a_3 + \frac{\varepsilon}{6y_c} \left| (a_1 + b_2) + (b_3 + 2a_4)x_c + (a_3 + 2b_5)y_c \right| \quad (3.42)$$

$$b_3^* = b_3 + \frac{\varepsilon}{6x_c} \left| (a_1 + b_2) + (b_3 + 2a_4)x_c + (a_3 + 2b_5)y_c \right| \quad (3.43)$$

$$a_4^* = a_4 + \frac{\varepsilon}{12x_c} \left| (a_1 + b_2) + (b_3 + 2a_4)x_c + (a_3 + 2b_5)y_c \right| \quad (3.44)$$

$$b_5^* = b_5 + \frac{\varepsilon}{12 y_c} \left| (a_1 + b_2) + (b_3 + 2a_4)x_c + (a_3 + 2b_5)y_c \right| \qquad (3.45)$$

and the vertical velocity is expressed as

$$\omega(p) = -\int_{p_0}^{p} \left\{ (a_1^* + b_2^*) + (b_3^* + 2a_4^*)x + (a_3^* + 2b_5^*)y \right\} dp \qquad (3.46)$$

The computation of the vertical velocity using the quadratic representation is not performed in this manual.

2. Vertical Velocity from Regularly Spaced Wind Data

If the wind observations are available on a regular grid array, the estimation of the kinematic vertical velocity is straightforward from the integration of the mass continuity equation as described by (3.11). The computational aspects are, however, somewhat complicated. Therefore, when estimating vertical motion, a single routine for the vertical integration should be consistently deployed each time. The horizontal divergence may be evaluated to a second or fourth order accuracy.

Two simple subroutines for the calculation of the vertical velocity using the kinematic method are provided for use in this section. Subroutine *KINOMGA* assumes the error in the divergence is proportional to its magnitude at all levels and has a second order accuracy in its finite difference estimates. Therefore, if the initial uncorrected divergence is of the same sign at all levels, this method fails, since the solution is $\omega = 0$ at all levels. This happens only when the data set at a particular grid point is of poor quality. In this case a simple four point smoother such as

$$\overline{D} = 0.25 \, (d_{i+1,j} + d_{i-1,j} + d_{i,j-1} + d_{i,j+1}) \qquad (3.47)$$

is recommended at these points where the uncorrected divergence has the same sign at all levels.

The second subroutine, *VMOTION*, uses a fourth order scheme and assumes the data to be more reliable at selected vertical levels. For example, at 1000 mb the surface and marine observations are available. Low level cloud motion provides a reasonable estimate of the wind field at 850 mb. Commercial aircraft and high cloud motion vectors generally cover the 300, 250, and 200 mb surfaces. Therefore, the divergence correction is only performed at the remaining levels. This is a superior routine for the estimation of the vertical motion and does not yield a zero solution even if the divergence has the same sign at all levels. Other subroutines invoked by *KINOMGA* and *VMOTION* are also provided and are fully documented.

3. Vertical Velocity from the Quasi-Geostrophic Omega Equation

The vertical velocity can also be obtained from the quasi-geostrophic form of the vorticity equation combined with the first law of thermodynamics. The quasi-geostrophic omega equation appears of appropriate use only over tropical regions where the vertical variation of vorticity advection and the thermal advection are important. Applications of this method can be made in situations of strong jet streams, and useful information can be obtained over tropical and subtropical regions. The use of this method is certainly most appropriate over the Asian monsoon region where substantial thermal advection is provided by the warm desert, and the combination of the westerly Somali jet together with the upper level tropical easterly jet constitutes an important source of differential vorticity advection.

The quasi-geostrophic omega equation may be obtained from the frictionless quasi-geostrophic vorticity equation and the adiabatic form of the first law of thermodynamics. The vorticity equation is expressed as

$$\frac{\partial \zeta}{\partial t} = -\vec{V} \cdot \nabla(\zeta + f) - \omega \frac{\partial \zeta}{\partial p} - (\zeta + f) \, \nabla \cdot \vec{V} + \vec{k} \cdot \left(\frac{\partial \vec{V}}{\partial p} \cdot \nabla \omega \right) + F_\zeta \quad (3.48)$$

and its frictionless quasi-geostrophic form may be written as

$$\frac{\partial \zeta}{\partial t} = -\vec{V} \cdot \nabla(\zeta + f) - f_0 \, \nabla \cdot \vec{V} \quad (3.49)$$

The geostrophic wind components are given by

$$u = -\frac{\partial \psi}{\partial y} \quad \text{and} \quad v = \frac{\partial \psi}{\partial x} \quad (3.50)$$

where ψ is a geostrophic streamfunction defined as

$$\psi = \frac{gz}{f_0} \tag{3.51}$$

The relative vorticity may then be expressed as

$$\zeta = \frac{\partial v}{\partial x} - \frac{\partial u}{\partial y} = \nabla^2 \psi \tag{3.52}$$

On the other hand, the adiabatic form of the thermodynamics equation gives

$$\frac{\partial T}{\partial t} = - \vec{V} \cdot \nabla T - \omega \frac{\partial T}{\partial p} + \frac{RT\omega}{C_p p} \tag{3.53}$$

Using the hydrostatic approximation, the temperature is expressed as a function of the streamfunction.

$$T = -g \frac{p}{R} \frac{\partial z}{\partial p} \tag{3.54}$$

or

$$T = - f_0 \frac{p}{R} \frac{\partial \psi}{\partial p} \tag{3.55}$$

Using this relation for temperature, (3.53) becomes

$$\frac{\partial}{\partial t} \left(\frac{\partial \psi}{\partial p} \right) = \vec{V} \cdot \nabla \left(\frac{\partial \psi}{\partial p} \right) + \sigma \frac{\omega}{f_0} \tag{3.56}$$

where σ is a measure of the static stability and is given by

$$\sigma = \frac{R}{p} \left[\frac{\partial T}{\partial p} - \frac{R}{C_p} \frac{T}{p} \right] \tag{3.57}$$

The derivative of (3.49) with respect to p and the Laplacian of (3.53) yield the so-called quasi-geostrophic omega equation,

$$\sigma \nabla^2 \omega + f_0^2 \frac{\partial^2 \omega}{\partial p^2} = f_0 \frac{\partial}{\partial p} J(\psi, \nabla^2 \psi + f) - f_0 \nabla^2 J(\psi, \frac{\partial \psi}{\partial p}) \tag{3.58}$$

In this equation, the only unknown is the vertical velocity. The streamfunction $\psi(x,y,p)$ is given at any instant of time. The forcing functions on the right hand side of (3.58) are named differential vorticity advection and Laplacian thermal advection, respectively. Equation (3.58) is an elliptic equation of the second order and requires boundary conditions along each of the six edges of the volume over which the vertical motion is to be estimated. For these homogeneous conditions, the quasi-geostrophic omega equation may be solved for each of the forcings separately. Therefore, the contribution to the vertical velocity from the differential vorticity advection is obtained as a solution of

$$\sigma \nabla^2 \omega_1 + f_0^2 \frac{\partial^2 \omega_1}{\partial p^2} = f_0 \frac{\partial}{\partial p} J(\psi, \nabla^2 \psi + f) \tag{3.59}$$

and

$$\sigma \nabla^2 \omega_2 + f_0^2 \frac{\partial^2 \omega_2}{\partial p^2} = - f_0 \nabla^2 J(\psi, \frac{\partial \psi}{\partial p}) \tag{3.60}$$

would provide the contribution to vertical motion from the thermal advection. The total vertical velocity is then obtained as

$$\omega = \omega_1 + \omega_2 \tag{3.61}$$

For a two-level model, it is convenient to express the vertical velocity as a parabolic function of pressure such as

$$\omega = ap^2 + bp + c \tag{3.62}$$

with the following boundary conditions:

$$\begin{cases} \omega = 0 & \text{at } p = 0 \text{ and } 1000 \text{ mb} \\ \omega = \omega_5 & \text{at } p = 500 \text{ mb} \end{cases} \tag{3.63}$$

where ω_5 represents the vertical velocity at 500 mb. In this case the vertical velocity at any level (p) may then be expressed as a function of ω_5. The functional relation is given by

$$\omega = -\omega_5 \left[\frac{p^2 - 1000p}{2.5 \ 10^5} \right] \tag{3.64}$$

With this parabolic assumption, the two level quasi-geostrophic omega equation reduces to a two dimensional Helmholtz equation of the form,

$$\sigma \nabla^2 \omega_5 - \frac{2f_0^2}{2.5\ 10^5}\ \omega_5 = F(x,y) \tag{3.65}$$

where $F(x,y)$ represents the two forcing functions. This equation is usually presented in a more convenient form as

$$\left[\nabla^2 - \frac{2f_0^2}{\sigma(2.5\ 10^5)} \right]\ \omega_5 = F'(x,y) \tag{3.66}$$

The static stability parameter, σ, is positive and constant in the two level model. Its value is about $0.02\ m^2s^{-2}mb^{-2}$ for the U.S standard atmosphere around 500 mb. The solution of this type of equation can be obtained numerically using either the relaxation method or the Fourier expansion which are presented in Chapter 4.

In the above development of the quasi-geostrophic omega equation, the effect of the condensation heating was omitted for simplicity. It is obvious that the introduction of the diabatic heating term will further complicate the equation. Nevertheless it is of considerable interest to examine its effect on the vertical motion in the context of the quasi-geostrophic theory.

If a dynamic ascent of saturated stable air is considered, the heating rate due to condensation is given by

$$H = -L\ \frac{dq_s}{dt} = -L\omega\ \frac{\partial q_s}{\partial p} \tag{3.67}$$

and is added as a diabatic forcing into the thermodynamics equation. L denotes the latent heat of condensation and q_s the saturation specific humidity. In the omega equation, the heating term will take the form

$$A_d = -\frac{R}{C_p p}\ \nabla^2 H \tag{3.68}$$

and the quasi-geostrophic omega equation for the saturated air is expressed as

$$\nabla^2 \left[\sigma - \frac{RL}{C_p p}\ \frac{\partial q_s}{\partial p} \right] \omega + f_0^2 \frac{\partial}{\partial p^2}\frac{\omega}{} = f_0 \frac{\partial}{\partial p}\ J(\psi, \nabla^2\psi + f) - f_0\ \nabla^2 J(\psi, \frac{\partial \psi}{\partial p}) \tag{3.69}$$

where the parameter,

$$\sigma_m = \sigma - \frac{RL}{C_p p} \frac{\partial q_s}{\partial p} \qquad (3.70)$$

is referred to as the moist static stability parameter.

The maximum limit of $\frac{\partial q_s}{\partial p}$ allowed within this framework poses a serious problem. In fact, the omega equation must remain elliptic if it is to be treated as a boundary value problem. This implies that σ_m must be positive definite and that the atmosphere be absolutely stable. The introduction of the heating term can render this equation nonelliptic, in which case a mathematical solution for the vertical velocity is not possible in the usual manner. In a conditionally unstable atmosphere, one should introduce some form of convective heating so that the ellipticity condition is preserved. For example, the omega equation can be decoupled by assuming the vertical motion to be proportional to the boundary layer heating below the condensation level.

Under the assumption that σ_m is positive, a solution to the omega equation for the moist case is obtained using the same procedure as for the dry case. Calculation of σ_m is carried out individually at all those points where $\omega < 0$ and the ratio q/q_s is close to unity. Elsewhere, where no large scale condensation occurs, the dry static stability parameter is used for the computation of omega. If the vertical velocity obtained in the dry case is subtracted from that obtained in the moist case, the resulting difference should represent the contribution from the heating due to condensation.

A complete Fortran code performing the calculation of the quasi-geostrophic vertical velocity for a two level model is provided. Program (*QG2LW*) computes the vertical velocity for both the dry and moist cases. The contribution to the vertical motion from the latent heat release is also calculated. This driver for the calculation of the vertical velocity uses subroutine **SIGMAL** to compute the moist static stability parameter and **RELAXW** to solve the Helmoltz equation. This later subroutine makes use of the relaxation method which is described in Chapter 4. The driver also uses the Laplacian and Jacobian solvers.

```
      program QG2LW
c
c     this program is used to solve the quasi-geostrophic two-level omega
c     equation incorporating subgrid scale stable condensation heating.
c
```

```
            parameter (l=21,m=13,np=7,l1=l-1,l2=l-2,m1=m-1,m2=m-2)
c
c     Declare variables.
c
            real bb(l,m),q(l,m),es(l,m),qs(l,m),phi(m),dx(l),cor(m)
            real omega(l,m),omegah(l,m),sigmah(l,m),a(l,m),b(l,m)
            real force(l,m),fzeta(l,m),ftheta(l,m),c(l,m),d(l,m)
            real datag (l,m,np),datat(l,m,np),datar(l,m,np)
            real z500(l,m),z1000(l,m),t750(l,m),rh750(l,m)
c
c     Define constants.
c
            data f0,g,alfa,eps/1.e-04,9.81,-1.6,.1e-08/
            data sigma,dy,dphi,dp/.02,2.5e05,2.5,500./
            pi         = 4.0*atan(1.0)
            phi(1)     = -15.
            c1         = pi/180.
c
c     Open input files .
c
            open (20,file='geop21.dat',status='old',readonly)
            open (30,file='temp21.dat',status='old',readonly)
            open (40,file='rel21.dat ',status='old',readonly)
c
c     Read multilevel input data
c
      15    format(10f8.2)
c
            do 3200 ip = 1, np
              read (20,15) ((datag(i,j,ip),i=1,l),j=1,m)
              read (30,15) ((datat(i,j,ip),i=1,l),j=1',m)
              read (40,15) ((datar(i,j,ip),i=1,l),j=1,m)
    3200    continue
            do 3202 i = 1, l
            do 3202 j = 1, m
c
c     Extract geopotential(meters) at 1000 and 500 mb .
c
              z1000 (i,j) = datag (i,j,1)
              z500  (i,j) = datag (i,j,4)
c
c     Extract temperature(kelvin) at 750 mb assumed to be =
c     (temp. at 700) + 2.
```

```
c
          t750 (i,j)  = datat(i,j,3) + 2.
c
c     extract relative humidity(%) at 750 mb assumed to be the same as
c     700 mb.
c
          rh750 (i,j) = datar(i,j,3)

 3202   continue
c
c     compute the grid distance and Coriolis parameter.
c
        do 3204 j = 2, m
 3204      phi(j) = phi(j-1) + dphi
        do 3206 j = 1, m
           dx(j)  = dy*cos(phi(j)*c1)
 3206   cor(j) = 2.*7.29/1.e05*sin(phi(j)*c1)
c
c     compute streamfunctions at level 1000 and 500 mb.for midlatitude
c     synoptic scales motions the streamfunction is approximated by:
c     psi = g*z/fo.
c
        do 3208 j = 1, m
        do 3208 i = 1, l
           z1000 (i,j) = z1000(i,j)*g/f0
           z500  (i,j) =  z500(i,j)*g/f0
 3208   continue
c
c        differential vorticity advection.
c     perform the laplacian of the streamfunction
c     and compute the jacobian of the streamfunction
c     and the absolute vorticity at both levels
c
          call LAP (a,z1000(1,1),dx,dy,l,m,l1,m1,l2,m2)
          call LAP (b,z500 (1,1),dx,dy,l,m,l1,m1.l2,m2)
c
        do 3210 j = 1, m
        do 3210 i = 1, l
           a(i,j) = a(i,j) + cor(j)
 3210      b(i,j) = b(i,j) + cor(j)
c
          call JAC (c,z1000(1,1),a,dx,dy,l,m,l1,m1,l2,m2)
          call JAC (d, z500(1,1),b,dx,dy,l,m,l1,m1,l2,m2)
```

```
c
c     Forcing due to differential vorticity advection
c
         do 3212 j = 1, m
         do 3212 i = 1, l
             sigmah (i,j) = sigma
 3212        fzeta (i,j) = (d(i,j)-c(i,j))/dp*f0
c
c           Laplacian of thermal advection.
c     performs the jacobian of streamfunction and vertical gradient
c     of streamfunction then form the laplacian
c
         do 3214 j = 1, m
         do 3214 i = 1, l
             c(i,j) = (z500 (i,j) - z1000(i,j))/dp
 3214        d(i,j) = (z1000(i,j) + z500 (i,j))*.5
c
         call JAC (a,d,c,dx,dy,l,m,l1,m1,l2,m2)
         call LAP (b,a,dx,dy,l,m,l1,m1,l2,m2)
c
c     Forcing due to laplacian of thermal advection
c
         do 3216 j = 1, m
         do 3216 i = 1, l
 3216    ftheta(i,j) = -f0*b(i,j)
c
c     Compute omega due to each forcing component
c     through relaxation method. Scale omega for display.
c
         call RELAXW (omega,fzeta,sigmah,l,m,dx,dy,l2,m2,eps.alfa)
c
         do 3218 i = 1, l
         do 3218 j = 1, m
 3218    omega (i,j) = omega(i,j)*1e+05
         print *,((omega(i,j),j=7,7),i=1,l,6)
         do 3220 j = 1,m
         do 3220 i = 1,l
             a(i,j) = omega(i,j)
 3220    continue
c
         call RELAXW (omega,ftheta,sigmah,l,m,dx,dy,l2,m2,eps.alfa)
c
         do 3222 i = 1, l
```

```
           do 3222 j = 1, m
    3222   omega(i,j) = omega(i,j)*1e+05
           print *,((omega(i,j),j=7,7),i=1,1,6)
  c
  c      Compute total forcing and omega (dry) due to both
  c      forcing contributions
  c
           do 3224 j = 1, m
           do 3224 i = 1, 1
              omega(i,j) = omega(i,j) + a(i,j)
    3224      force(i,j) = fzeta(i,j) + ftheta(i,j)
           do 3226 j = 1, m
           do 3226 i = 1, 1
              if ( omega(i,j).ge.0. ) go to 70
  c
  c      Stable heating is invoked where rising motion is present,
  c      the modified stability parameter is computed by sigmal.
  c
           call SIGMAL (sigmah,t750,rh750,sigma,l,m,bb,q,es,qs)
           go to 3226
     70    sigmah(i,j) = sigma
    3226   continue
  c
  c      Compute omega for the moist case.
  c      omega moist is the difference between the total
  c      omega with heating and the dry omega.
  c
           call RELAXW (omegah,force,sigmah,l,m,dx,dy,l2,m2.eps,alfa)
  c
  c      Scale and display omega
  c
           do 3228 i = 1, 1
           do 3228 j = 1, m
    3228   omegah(i,j) = omegah(i,j)*1e+05
           print *,((omegah(i,j),j=7,7),i=1,1,6)
  c
           do 3230 j = 1, m
           do 3230 i = 1, 1
              omegah(i,j) = omegah(i,j) - omega(i,j)
    3230   continue
           stop
           end
```

The input fields for this driver are the geopotential height at 1000 and 500 mb together with the temperature and relative humidity in the middle of the layer (1000-500 mb). These are shown in Figures (3.1) and (3.2), respectively.

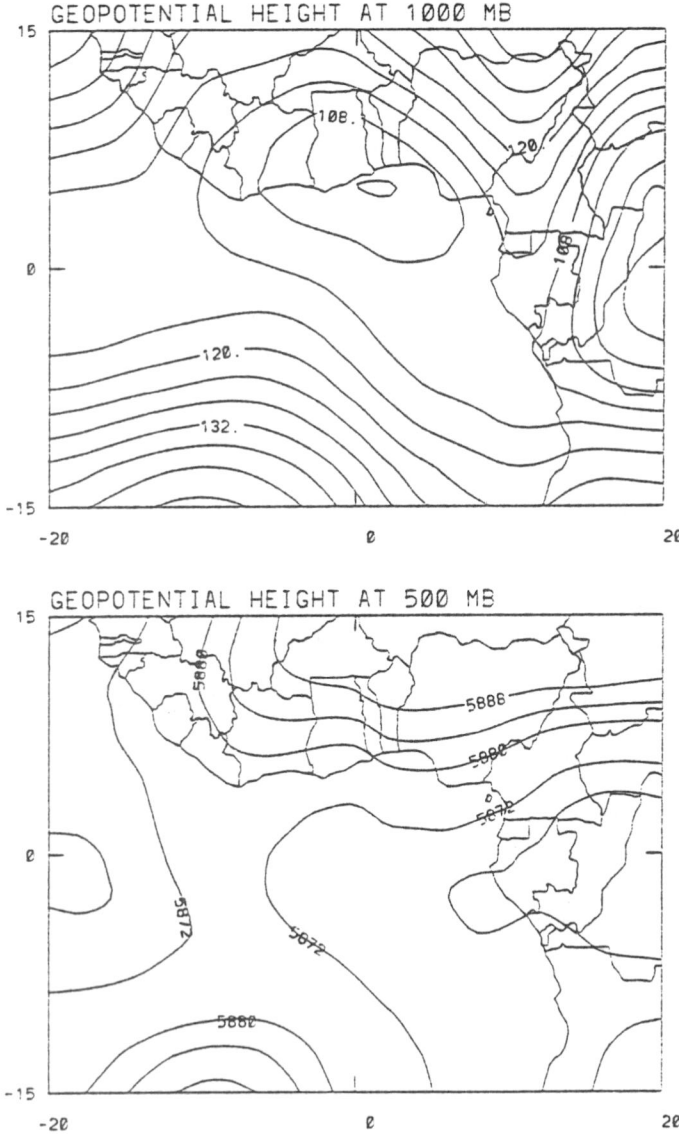

Figure 3.1: Geopotential height at 1000 mb and 500 mb (meters).

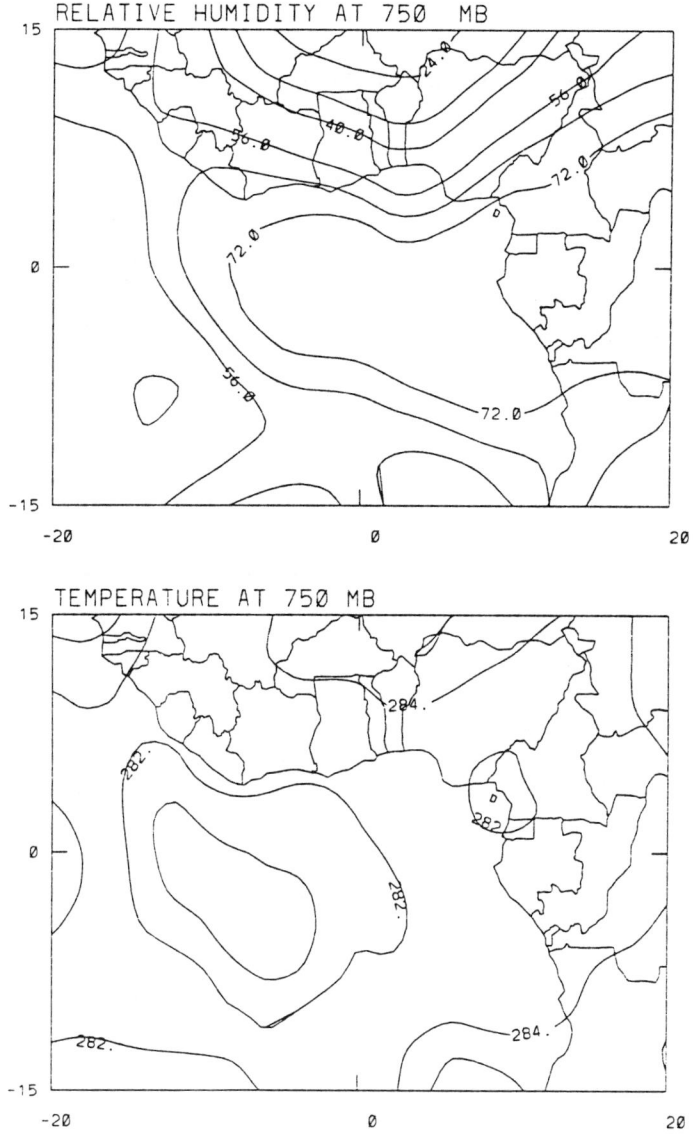

Figure 3.2: Relative humidity (percent) and temperature at 750 mb (°K).

The estimated contributions to omega from the thermal advection and the differential vorticity advection are presented in Figure (3.3).

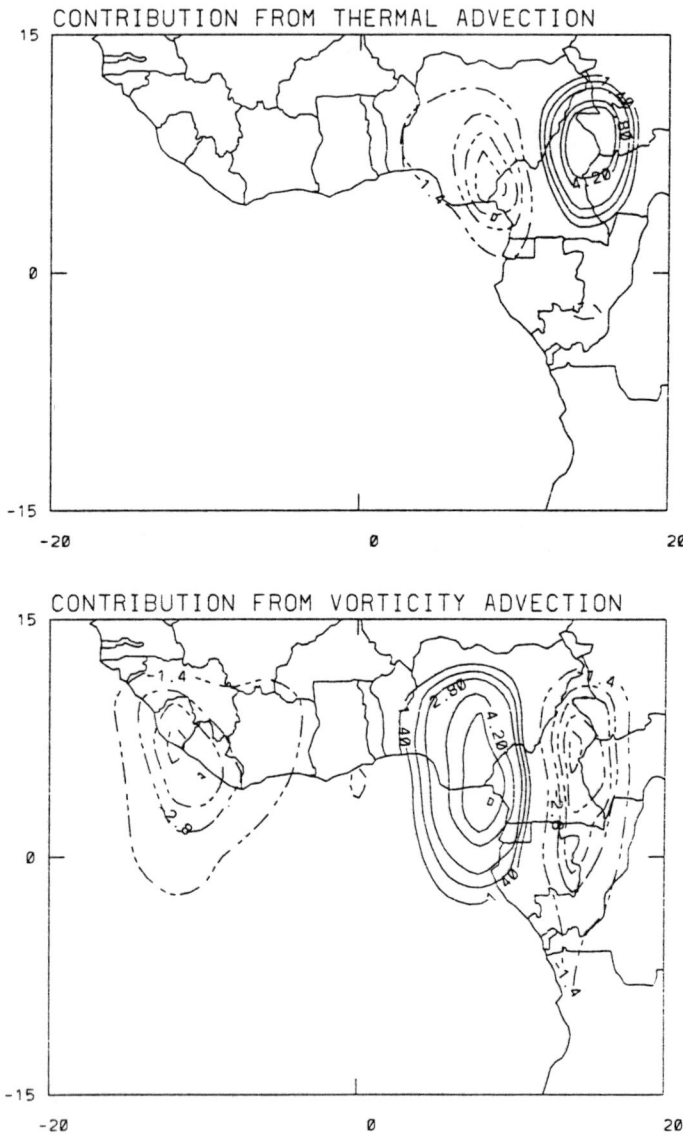

Figure 3.3: Contribution to omega from thermal advection and
            vorticity advection (10^{-5}mbs^{-1}). (Dashed values are
            negative and correspond to upward motion.)

The dry and moist situation total vertical velocity is illustrated in Figure (3.4), and the condensational heating contribution is mapped in Figure (3.5).

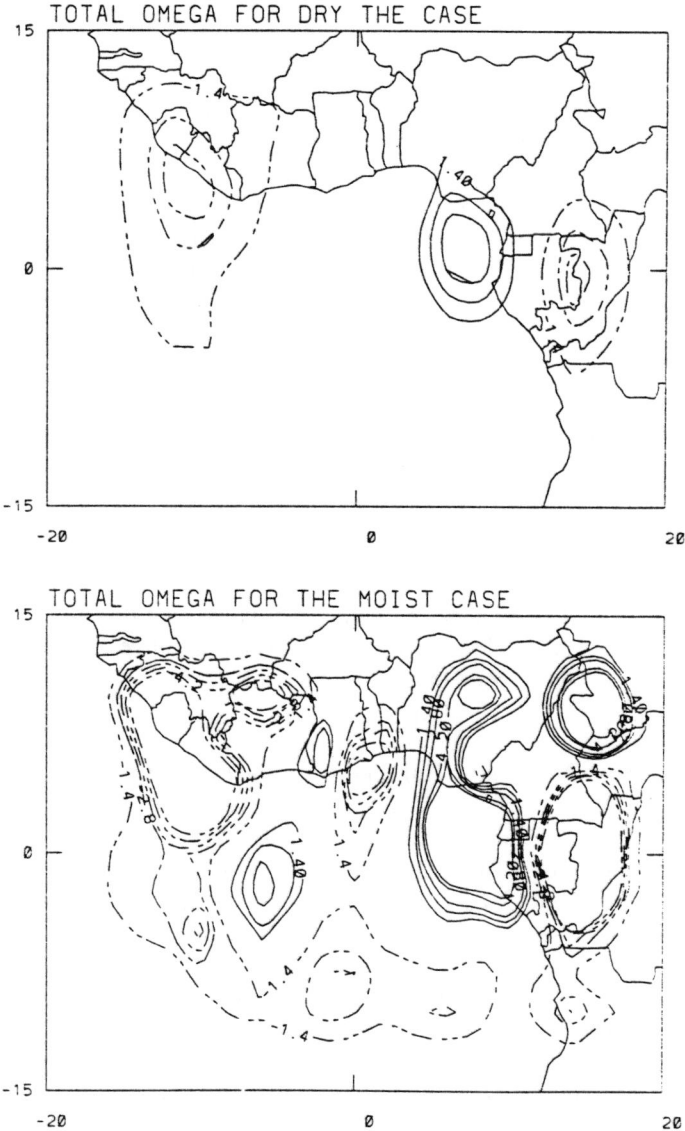

Figure 3.4:   Total omega for the dry and the moist cases
              (10^{-5}mbs^{-1}). (Dashed values are negative and corres-
              pond to upward motion.)

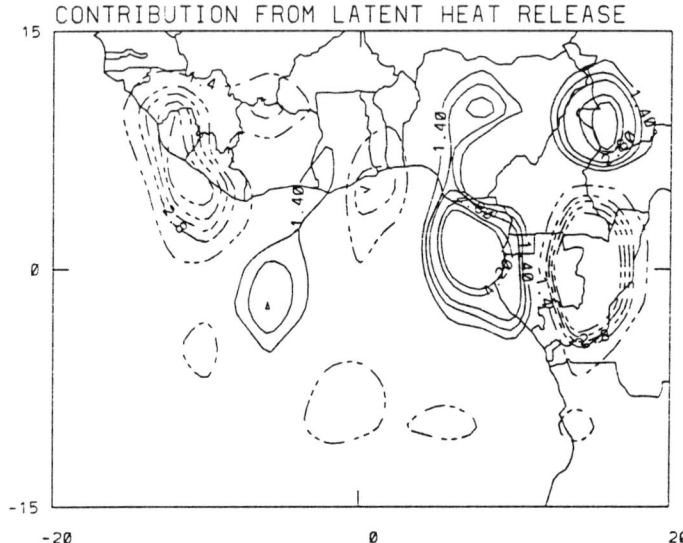

CONTRIBUTION FROM LATENT HEAT RELEASE

Figure 3.5: Contribution to omega from latent heat release
(10^{-5}mbs^{-1}). (Dashed values are negative and
correspond to upward motion.)

4. Multilevel Nonlinear Balance Omega Equation

The last method for the calculation of the vertical motion described
in this chapter is derived from a multilevel nonlinear balance equation.
This powerful method may be applied with considerable success to
subtropical disturbances. The mathematical framework of the multilevel
nonlinear balance model appears as a normal extension to the
quasi-geostrophic model. However, unlike the quasi-geostrophic, this
model includes the effects of air-sea interaction, convective heating,
stable heating, radiative heating and friction.

In a pressure coordinate system, the quasi-static equation of motion
may be written as

$$\frac{\partial \vec{V}}{\partial t} + \vec{V} \cdot \nabla \vec{V} + \omega \frac{\partial \vec{V}}{\partial p} - f\vec{V} \times \vec{k} = -g\nabla z + \vec{F} \tag{3.71}$$

and the hydrostatic equation is expressed in the form,

$$\frac{R\theta}{p} \left[\frac{p}{p_0} \right]^{R/c_p} = -g \frac{\partial z}{\partial p} \qquad (3.72)$$

The continuity equation is given by

$$\nabla \cdot \vec{V} = -\frac{\partial \omega}{\partial p} \qquad (3.73)$$

and the thermodynamics equation is expressed as

$$C_p \frac{T}{\theta} \left[\frac{\partial \theta}{\partial t} + \vec{V} \cdot \nabla \theta + \omega \frac{\partial \theta}{\partial p} \right] = H \qquad (3.74)$$

where H represents the total diabatic heating per unit mass of air. In the formulation of a diagnostic balance model, the vorticity and divergence equations are obtained from (3.71) and (3.72). For scale analysis considerations, the time derivative of the divergence and the advection by the divergent part of the wind are generally neglected, Phillips (1963).

Expressing the wind field as a function of the streamfunction ψ and the velocity potential χ,

$$\vec{V} = \vec{k} \times \nabla \psi - \nabla \chi \qquad (3.75)$$

the vorticity and divergence equations may be, respectively, written as

$$\frac{\partial}{\partial t} \nabla^2 \psi = -J(\psi,\xi_a) + \nabla \chi \cdot \nabla \xi_a + \xi_a \nabla^2 \chi - \omega \frac{\partial}{\partial p} \nabla^2 \psi - \nabla \omega \cdot \nabla \frac{\partial \psi}{\partial p}$$

$$- g \frac{\partial}{\partial p} \left[\frac{\partial \tau_y}{\partial x} - \frac{\partial \tau_x}{\partial y} \right] \qquad (3.76)$$

$$\nabla \cdot f \nabla \psi = \nabla^2 \phi - 2J \left(\frac{\partial \psi}{\partial x}, \frac{\partial \psi}{\partial y} \right) \qquad (3.77)$$

The frictional force is retained at 1000 mb surface and is defined in terms of stresses,

$$\tau_x = \rho \, C_d \, u \, \sqrt{u^2 + v^2} \qquad (3.78)$$

$$\tau_y = \rho \, C_d \, v \, \sqrt{u^2 + v^2} \qquad (3.79)$$

where u and v are the horizontal wind components on a pressure surface, and C_d is the drag coefficient. For low latitudes systems, the relative vorticity is obtained from the horizontal wind analysis,

$$\nabla^2 \psi = \frac{\partial v_0}{\partial x} - \frac{\partial u_0}{\partial y} \qquad (3.80)$$

where u_0 and v_0 are the zonal and meridional components of the observed wind, and the streamfunction is assumed to be related to the geopotential height through the balance equation,

$$\nabla^2 \phi = \nabla . f \nabla \psi + 2J\left(\frac{\partial \psi}{\partial x}, \frac{\partial \psi}{\partial y}\right) \qquad (3.81)$$

On the other hand, with the introduction of

$$\pi = \frac{RT}{p\,\theta} \qquad (3.82)$$

the thermodynamic energy equation may be expressed as

$$\pi \frac{\partial \theta}{\partial t} = - \pi \, J(\psi,\theta) + \pi \, \nabla\chi.\nabla\theta + \sigma\omega + \frac{HR}{C_p p} \qquad (3.83)$$

where σ is the stability parameter and is given by

$$\sigma = - \pi \frac{\partial \theta}{\partial p} \qquad (3.84)$$

Finally, the general balance model is obtained by combining (3.76), (3.77) and (3.83). It is governed by the following three equations for ω, χ and $\frac{\partial \psi}{\partial t}$.

$$\nabla^2 \sigma\omega + f^2 \frac{\partial^2 \omega}{\partial p^2} = f \frac{\partial}{\partial p} J(\psi,\xi_a) + \pi \nabla^2 J(\psi,\theta) - 2 \frac{\partial}{\partial t} \frac{\partial}{\partial p} J\left(\frac{\partial \psi}{\partial x}, \frac{\partial \psi}{\partial y}\right)$$

$$- f \frac{\partial}{\partial p} (\xi \nabla^2 \chi) + f \frac{\partial}{\partial p} g \frac{\partial}{\partial p} (\frac{\partial \tau_y}{\partial x} - \frac{\partial \tau_x}{\partial y}) - \frac{R}{C_p p} \nabla^2 H$$

$$- \frac{R}{C_p p} \nabla^2 H_s + f \frac{\partial}{\partial p} (\omega \frac{\partial}{\partial p} \nabla^2 \psi) + f \frac{\partial}{\partial p} (\nabla \omega. \nabla \frac{\partial \psi}{\partial p})$$

$$- f \frac{\partial}{\partial p} (\nabla \chi. \nabla \xi_a) - \pi \nabla^2 (\nabla \chi. \nabla \theta) - \beta \frac{\partial}{\partial p} \frac{\partial}{\partial y} \frac{\partial \psi}{\partial t} \quad (3.85)$$

$$\nabla^2 \chi = \frac{\partial \omega}{\partial p} \quad (3.86)$$

$$\nabla^2 \frac{\partial \psi}{\partial t} = - J(\psi, \xi_a) - g \frac{\partial}{\partial p} (\frac{\partial \tau_y}{\partial x} - \frac{\partial \tau_y}{\partial y}) + \nabla \chi. \nabla \xi_a + \xi_a \nabla^2 \chi - \nabla \omega. \nabla \frac{\partial \psi}{\partial p}$$

$$- \omega \frac{\partial}{\partial p} \nabla^2 \psi \quad (3.87)$$

The numerical solution of this system poses many important questions. Among these, one can mention the boundary conditions for ω, χ and $\frac{\partial \psi}{\partial t}$, the formulation of the heating terms for the stable and unstable conditions, the inclusion of the terrain and the surface friction effects. The ellipticity of the three equations and the validity of the boundary value technique need also to be carefully examined in conjunction with the diabatic terms.

4.1 The Development Problem

In the quasi-geostrophic model, provided the geopotential distribution is known, the vertical motion can be obtained by solving the omega equation. In the case these velocities are used to construct the three-dimensional trajectories, the change in vorticity following the parcel may be related to the computed convergence of mass,

$$\frac{d\xi_a}{dt} = f_o \frac{\partial \omega}{\partial p} \quad (3.88)$$

Furthermore, the map features which contribute to vertical motions and convergence may also be determined. In this context, the development criteria introduced by Sutcliffe (1947) and Petterssen (1956) appear to be based on a quasi-geostrophic formulation that pursue similar reasoning. Conversely, in a general balance model this question gives rise to many complications. In this case the horizontal wind components are obtained from the streamfunction and potential velocity as

$$u = -\frac{\partial \chi}{\partial x} - \frac{\partial \psi}{\partial y} \tag{3.89}$$

$$v = -\frac{\partial \chi}{\partial x} + \frac{\partial \psi}{\partial y} \tag{3.90}$$

and the vertical velocity for several map times may be used to construct the three-dimensional trajectories, Paegle (1966). Along such trajectories, various development terms of the vorticity equation are involved, and various baroclinic mechanisms inducing convergence and rising motions can be listed. This information becomes of capital importance when investigating storm developments.

4.2 Forcing Functions of the Balance Omega Equation

Unlike the quasi-geostrophic omega equation which is controlled only by two forcings, the general balance model has 12 forcing functions. At a first sight it is not obvious that an analysis should be carried this far; however, the contribution from each of these terms is of significant importance, in that it brings information that is not obtainable from the quasi-geostrophic model.

1.	$f \dfrac{\partial}{\partial p} J(\psi, \zeta_a)$	Differential vorticity advection by the nondivergent part of the wind.
2.	$\pi \nabla^2 J(\psi, \theta)$	Laplacian of thermal advection by the nondivergent part of the wind.
3.	$-2 \dfrac{\partial}{\partial t} \dfrac{\partial}{\partial p} J(\dfrac{\partial \psi}{\partial x}, \dfrac{\partial \psi}{\partial y})$	Differential deformation effect.
4.	$-f \dfrac{\partial}{\partial p} (\zeta \nabla^2 \chi)$	Differential divergence effect.

5. $\quad f \dfrac{\partial}{\partial p} g \dfrac{\partial}{\partial p} \left(\dfrac{\partial \tau_y}{\partial x} - \dfrac{\partial \tau_x}{\partial y} \right)$ 　　Effect of frictional stresses.

6. $\quad -\dfrac{R}{c_p p} \nabla^2 H_L$ 　　Effect of latent heat release.

7. $\quad -\dfrac{R}{c_p p} \nabla^2 H_s$ 　　Effect of sensible heat transfer.

8. $\quad f \dfrac{\partial}{\partial p} \left(\omega \dfrac{\partial}{\partial p} \nabla^2 \psi \right)$ 　　Differential vertical advection of vorticity

9. $\quad f \dfrac{\partial}{\partial p} \left(\nabla \omega \cdot \nabla \dfrac{\partial \psi}{\partial p} \right)$ 　　Differential turning of vortex tubes.

10. $\quad -f \dfrac{\partial}{\partial p} (\nabla \chi \cdot \nabla \zeta_a)$ 　　Differential advection of vorticity by the divergent part of the wind.

11. $\quad -\pi \nabla^2 (\nabla \chi \cdot \nabla \theta)$ 　　Laplacian of thermal advection by the divergent part of the wind.

12. $\quad -\beta \dfrac{\partial}{\partial p} \dfrac{\partial}{\partial y} \dfrac{\partial \psi}{\partial t}$ 　　Contribution by the beta term of the vorticity equation.

If a forcing function is positive it leads to upward motion generally, and the converse also holds. Thus positive vorticity aloft and localized strong warm advection are good candidates for rising motion. The other terms are somewhat more difficult to interpret. However, if they have a localized strong positive value, they tend to be regions of strong upward motion. The frictional contribution tends to produce upward motion where the surface vorticity is cyclonic. The heating terms tend to produce upward motion over regions of net strong positive heating, and the converse also holds (Krishnamurti, 1968).

The complete problem may be solved with or without terrain effects. It is however desirable to include the terrain influence as an internal forcing. Furthermore, the pressure coordinate system is somewhat artificial near the surface due to data reduction to sea level. This boundary condition constitutes only a compromise for the real air motion along the sloping topography. An earth frame, where the ground level is a coordinate surface, would be more appropriate for the introduction of the terrain as a boundary condition. It should be noted, however, that the problem becomes more complicated as all the terms of the balance model are retained. Nevertheless, the interpretation of all the forcing terms in the nonlinear balance omega equation appears to follow a general qualitative rule as far as the contribution to vertical motion is concerned. Though some exceptions exist, a positive forcing function contributes generally to the production of rising motions. The relation between the vertical motion, the vorticity and thermal advection patterns appears then relatively easy to assess. Synoptic experience and simple quasi-geostrophic omega equation solutions have, however, shown an inverse relationship between these forcing functions and the sign of the vertical motion. It should be noted that the first two forcing functions are not the same within the two models. Unlike the first model, where the stream-function is geostrophic, it is the balance nondivergent streamfunction which is used in the second model. Furthermore, in the quasi-geostrophic theory, f_0 is used instead of f and the static stability, σ, is function of pressure only, while in the balance model it is a tridimensional variable of x, y and p. Similarly, a qualitative interpretation can be made for terms 1,2,8,9,10 and 11. Terms 6 and 7 will be positive if H_L and H_S are

positive and will in general contribute rising motions. The heating function is described as

$$F_H = -\frac{R}{C_p p} \nabla^2 H \tag{3.91}$$

where H is the total diabatic heating rate per unit mass of air defined through the first law of thermodynamics as

$$C_p \frac{T}{\theta} \frac{d\theta}{dt} = H \tag{3.92}$$

The units of the heating function are $m^2 s^{-3}$. Treatments of the diabatic physical processes such as convection, large scale condensation and radiative heating are discussed in appropriate sections of this book. The effect of frictional stress is to enhance upward vertical motion in cyclonic relative vorticity regions and sinking motion in regions dominated by anticyclonic relative vorticity. The frictional stresses can be defined using simple bulk aerodynamic relations such as

$$\tau_x = \rho \, C_d \, u \, \sqrt{u^2 + v^2} \tag{3.93}$$

and

$$\tau_y = \rho \, C_d \, v \, \sqrt{u^2 + v^2} \tag{3.94}$$

The stresses are assumed to vanish at the top of the planetary boundary layer. Further refinements in the definition of the stresses components can be made.

The deformation and divergence terms which are inexistent in the quasi-geostrophic formulation appear to be important, and their interpretation in the balance model is somewhat difficult. Let D_1 and D_2 be the components of the deformation field such as

$$D_1 = \frac{\partial u_\psi}{\partial x} - \frac{\partial v_\psi}{\partial y} \tag{3.95}$$

and

$$D_2 = \frac{\partial v_\psi}{\partial x} - \frac{\partial u_\psi}{\partial y} \tag{3.96}$$

It can then be shown that

$$2D_1^2 + D_2^2 - \zeta^2 = -4J(u_\psi, v_\psi) \tag{3.97}$$

In regions of intensifying frontogenesis, the magnitude of the deformation generally increases and the contribution to the vertical motion by the term $-\dfrac{\partial}{\partial t}\dfrac{\partial}{\partial p}\, J(u_\psi, v_\psi)$ is expected to be significant. The divergence term acts to alter the vertical motion distribution produced by the leading two terms in regions of large relative vorticity. This is directly visible considering the form of the forcing function, $-f\dfrac{\partial}{\partial p}(\zeta \nabla^2 \chi)$. Assume positive relative vorticity in regions of strong sinking motion. Therefore, at low levels $\zeta \nabla^2 \chi < 0$ and $-f\dfrac{\partial}{\partial p}(\zeta \nabla^2 \chi)$ is positive, contributing then to rising motion and opposing the two leading terms. The converse is true in regions of upward motion. The β term (12) appears to be the least important of all contributions.

4.3 The Boundary Conditions

Various boundary conditions are to be set when solving the multilevel nonlinear balance omega equation. Along the zonal direction, cyclic boundary conditions are used for all variables. This technique is addressed in Chapter 4. The boundary conditions for the determination of the streamfunction from the wind field and the geopotential from the streamfunction are also discussed later. For the velocity potential, the boundary conditions are such that

$$\chi = 0 \quad \text{at} \quad \begin{cases} y = y_1 \\ y = y_2 \end{cases} \tag{3.98}$$

where y_1 and y_2 are the meridional limits of the domain. Similar conditions are imposed on the streamfunction tendency at these boundaries. Since the omega equation is three dimensional, it requires boundary conditions along the zonal, meridional and vertical axes. The cyclic condition is used zonally, and omega is set to zero at the meridional boundaries. Omega is also set to zero at the top of the model atmosphere, while at the ground level it is either set to zero or induced from the terrain field using the relation

$$\omega_0 = - \frac{g P_0}{R T_0} (\vec{V}_0 \cdot \nabla z_0) \tag{3.99}$$

where $\vec{V}_0$ is the horizontal wind, z_0 the terrain elevation, and P_0, T_0 represent the pressure and temperature at the surface, respectively.

5. Numerical Algorithms

A complete Fortran code giving the solution of the nonlinear balance omega equation is not provided in this manual. However, a list of the major numerical algorithms needed to solve the system is presented and discussed in this section.

i)   Subroutine **RELAXT** is a two dimensional Poisson solver. It can be used to obtain the streamfunction from the wind field using

$$\nabla^2 \psi = \frac{\partial v}{\partial x} - \frac{\partial u}{\partial y} \qquad (3.100)$$

or the velocity potential from the vertical motion,

$$\nabla^2 \chi = \frac{\partial \omega}{\partial p} \qquad (3.101)$$

It would also be used to solve for the geopotential from the balance equation and for the streamfunction tendency from the vorticity equation. The method of solution for the Poisson type of equation is carried out using the standard relaxation method which is described in Chapter 4.

ii)   Subroutine **ROME** is a three dimensional Poisson solver and would be primarily used to solve the omega equation, i.e.,

$$\nabla^2 \sigma \omega + f^2 \frac{\partial^2 \omega}{\partial p^2} = F(x,y,z) \qquad (3.102)$$

For the complete solution of the nonlinear balance omega equation, many other subprograms are needed to compute the following quantities:

3.   Jacobian
4.   Laplacian
5.   Convective heating
6.   Large scale condensation heating
7.   Air-sea interaction forcing
8.   Frictional stress
9.   Radiative heating

All these routines are available within this book and are not repeated here. A procedure for the solution of the omega equation is, however, suggested as follows:

1.   Using u and v at different levels, obtain the streamfunction (3.52).
2.   Obtain the geopotential from the streamfunction (3.81).
3.   Obtain the quasi-geostrophic omega (3.58).
4.   Obtain the velocity potential from the continuity equation (3.86).
5.   The streamfunction tendency is computed from the vorticity equation (3.87).

6.   Finally, solve the complete omega equation using the three-dimensional Poisson solver (3.85).

7.   Iterate steps 4, 5, and 6 at least four times for convergence.

4

Estimation of Streamfunctions, Velocity Potential, and Geopotential Height from the Wind Field

Wind derived quantities such as streamfunctions and velocity potential have introduced a great simplicity in the expression of the governing equations of the atmospheric motion and constitute the basic predictors for many numerical weather prediction models. It is, therefore, desirable to develop a transform pair which computes these potential functions given the observed wind field. The inverse transform would provide the wind field from the streamfunctions and velocity potential.

Following the two dimensional Helmoltz theorem, the horizontal wind may be expressed as a function of its rotational and divergent components as

$$\vec{V} = \vec{k} \times \nabla \psi - \nabla \chi \qquad (4.1)$$

where ψ is the streamfunction and χ is the velocity potential. The scalar equations for the horizontal wind components over a finite two dimensional domain can then be written as

$$u = -\frac{\partial \psi}{\partial y} - \frac{\partial \chi}{\partial x} \qquad (4.2)$$

and

$$v = \frac{\partial \psi}{\partial x} - \frac{\partial \chi}{\partial y} \qquad (4.3)$$

It is easy to see that from (4.2) and (4.3) one can obtain

$$\nabla^2 \psi = \frac{\partial v}{\partial x} - \frac{\partial u}{\partial y} \qquad (4.4)$$

and

73

$$\nabla^2\chi = -\left[\frac{\partial u}{\partial x} + \frac{\partial v}{\partial y}\right] \tag{4.5}$$

which may be rewritten as

$$\nabla^2\psi = \xi \tag{4.6}$$

and

$$\nabla^2\chi = -D \tag{4.7}$$

where ξ and D represent the relative vorticity and horizontal divergence, respectively.

Equations (4.2) and (4.3) represent the inverse transform which reconstitute the wind field from the streamfunction and velocity potential. This operation is quite simple and has been discussed in Chapter 2. The direct transform, computing the streamfunction and velocity potential from the wind field, is obtained via the solution of Helmoltz equations (4.6) and (4.7) subject to appropriate boundary conditions.

In this chapter, two powerful and commonly used methods for the obtention of the streamfunction and velocity potential from the analyzed wind field are proposed. The wind field is assumed to be known on a regularly spaced grid array.

1. Relaxation Method

Given the horizontal wind components, the vorticity and divergence may be calculated as

$$\xi = \frac{\partial v}{\partial x} - \frac{\partial u}{\partial y} \tag{4.8}$$

$$D = \frac{\partial u}{\partial x} + \frac{\partial v}{\partial y} \tag{4.9}$$

and the streamfunctions and velocity potential can be obtained as solutions to Poisson equations (4.6) and (4.7). Boundary conditions represent a major problem for the solutions of these equations, especially for the determination of the streamfunction. In this case, the assumption that the outward normal velocity at the boundary can be corrected to yield a net zero outward mass flux is made. This is expressed as

$$\oint (V_n)_c \, ds = 0 \qquad (4.10)$$

and

$$(V_n)_c = V_n + \varepsilon \, |V|_n \qquad (4.11)$$

where $\oint$ means the integral is carried out along the contour limiting the domain and ds is a length element along this contour. Equation (4.10) assumes the correction of the outward normal wind is proportional to its magnitude and ε is the correction factor. At the boundaries the normal wind is given by

$V_n = -V$	along southern boundary
$V_n = -U$	along western boundary
$V_n = +V$	along northern boundary
$V_n = +U$	along eastern boundary

This technique constitutes a reasonable boundary condition provided that the domain is large enough to support the mass flux conservation assumption. It is appropriate to use domains with at least 60 degrees of longitude in the zonal direction and 40 degrees of latitude in the meridional direction. The boundary wind components are corrected at each point using (4.10) and the streamfunction is assumed to be known at the northwestern corner of the domain, eg., $\psi = 0$ at $i = 1$, $j = M$. Its value at the other points along the boundary is calculated using the corrected normal velocity as

$$\frac{\partial \psi}{\partial s} = (V_n)_c \qquad (4.12)$$

or

$$\psi_2 = \psi_1 + \frac{(V_{n1})_c + (V_{n2})_c}{2} \Delta s \qquad (4.13)$$

where Δs is the grid distance along the considered boundary.

The next step consists of solving Poisson equation $\nabla^2 \psi = \xi(x,y)$ subject to these boundary conditions. In this case, the standard over-relaxation technique is used. Basically the relaxation procedure is carried out in two steps using an iterative scheme over the entire domain. First, the residual between the forcing function and the finite difference analog value of the Laplacian of the streamfunction is evaluated as

$$R = \nabla^2 \psi - \xi \qquad (4.14)$$

next the streamfunction is estimated as

$$\psi^{v+1} = \psi^v + \alpha R \sqrt{\Delta x \Delta y} \qquad (4.15)$$

where α is the over-relaxation factor and Δx, Δy are the grid spacing along x and y directions, respectively. Usually, with a value of α ranging between 0.2 and 0.5, this method appears to be appropriate for most problems. However, the optimal value of α for a grid size of $2°$ latitude/longitude is around 0.47. An initial guess field of the streamfunction is necessary for the iterative process of the relaxation. The guess field can, in principle, be specified as zero everywhere except at the boundaries. The order of magnitude of ψ is about 10^7 m^{2}s^{-1} and $\Delta \psi$ is of the order of 10^6 m^{2}s^{-1}. If $\Delta x = 10^5$ m, $\Delta \psi / \Delta x$ is approximately 10 ms^{-1}. Therefore, an appropriate margin of tolerance for ψ would be between 10^3 and 10^4 m^{2}s^{-1}. The symbol v represents the iteration count and is incremented until the residual R is less than some specified precision factor.

The calculation of the velocity potential is done by solving

$$\nabla^2 \chi = - \left[\frac{\partial u}{\partial x} + \frac{\partial v}{\partial y} \right] \qquad (4.16)$$

where the divergence is supposed to be known. The choice of $\chi = 0$ at the boundaries is reasonable if the domain is sufficiently large. The solution of this problem is generally carried out adequately with a velocity potential equal to zero as a guess field. The tolerance error should be of the same order of magnitude as that allowed for the streamfunction.

A program which reads the horizontal wind components and makes use of the over-relaxation technique to estimate the streamfunctions is presented at the end of Section 2. The initial wind field used to carry out these calculations is shown in the form of streamlines (Figure 4.1).

2. Fourier Transform Method

In this method, the solutions for Poisson's equations are based on the double Fourier expansions assuming periodic boundary conditions for a limited domain. The calculations are consistent with the following scheme adopted by Stephens and Johnson (1978).

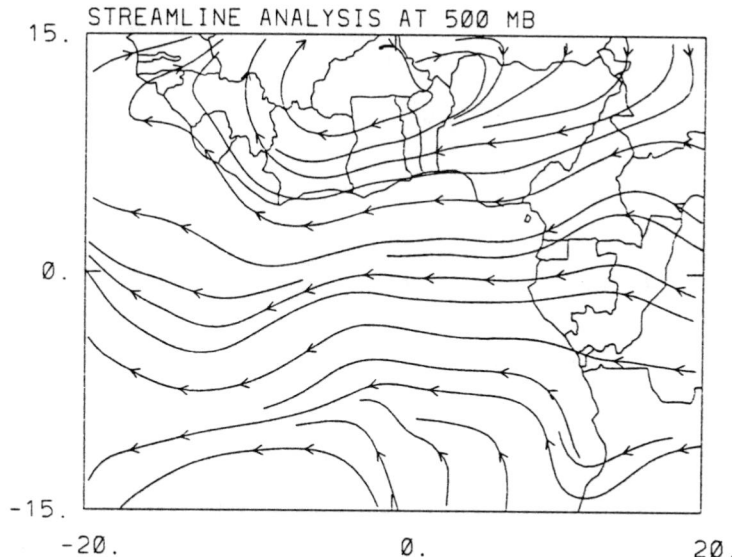

Figure 4.1: Streamlines of the initial wind field.

$$u_{jk} = u(j\Delta x, k\Delta y) \tag{4.17}$$

$j = 0, \dots , L-1$ and $k = 0, \dots , M-1$ for an $L \times M$ domain.

Thus,

$$u_{jk} = -\left[\frac{\Psi_{j,k+1} - \Psi_{j,k}}{\Delta y}\right] - \left[\frac{\chi_{j,k} - \chi_{j-1,k}}{\Delta x}\right] \tag{4.18}$$

and

$$v_{jk} = \left[\frac{\Psi_{j+1,k} - \Psi_{j,k}}{\Delta x}\right] - \left[\frac{\chi_{j,k} - \chi_{j,k-1}}{\Delta y}\right] \tag{4.19}$$

If Δ_x and ∇_x represent the forward and backward difference operator along the x direction, respectively, then the above equations can be written as

$$u_{jk} = -\Delta_y \Psi_{jk} - \nabla_x \chi_{jk} \tag{4.20}$$

and

$$v_{jk} = \Delta_x \Psi_{jk} - \nabla_y \chi_{jk} \tag{4.21}$$

Differentiating equations (4.20) with respect to y and (4.21) with respect to x using forward schemes one obtains

$$\nabla_y u_{jk} = - \nabla_y \Delta_y \Psi_{jk} - \nabla_y \nabla_x \chi_{jk} \tag{4.22}$$

$$\nabla_x v_{jk} = \nabla_x \Delta_x \Psi_{jk} - \nabla_x \nabla_y \chi_{jk} \tag{4.23}$$

Combining (4.22) and (4.23) a difference analog is obtained for Equation (4.4) that is,

$$\left[\nabla_x \Delta_x + \nabla_y \Delta_y \right] \Psi_{jk} = \nabla_x v_{jk} - \nabla_y u_{jk} \tag{4.24}$$

While u and v can be cyclic over the domain, Ψ and χ cannot be cyclic over the domain when the area averaged $\bar{u}$ and $\bar{v}$ do not vanish. Therefore, the contributions from $\bar{u}$ and $\bar{v}$ are treated separately. If Ψ and χ are cyclic along the x-direction then their zonal wind contribution U_χ averaged over the domain is zero. The same is true for V_ψ. A similar problem arises if there is a cyclic condition along the meridional direction for the components U_ψ and V_χ. It is for this reason that one treats $\bar{u}$ and $\bar{v}$ separately. They are included in Ψ and χ fields as $\Psi \equiv \tilde{\Psi} + \Psi'$ and $\chi \equiv \tilde{\chi} + \chi'$ such that $\tilde{\Psi}_k = - \bar{u}k\Delta y$ and $\tilde{\chi}_k = - \bar{v}k\Delta y$.
From (4.20) and (4.21) one gets

$$u_{jk} - \bar{u} = - \Delta_y \Psi'_{jk} - \nabla_x \chi'_{jk} \tag{4.25}$$

and

$$v_{jk} - \bar{v} = \Delta_x \Psi'_{jk} - \nabla_y \chi'_{jk} \tag{4.26}$$

Following the same steps as for (4.22) and (4.23), difference analog equations are obtained for the perturbations Ψ' and χ'.

$$\left[\nabla_x \Delta_x + \nabla_y \Delta_y \right] \Psi'_{jk} = \nabla_x v_{jk} - \nabla_y u_{jk} \tag{4.27}$$

$$\left[\nabla_x \Delta_x + \nabla_y \Delta_y\right]\chi'_{jk} = -\Delta_x u_{jk} - \Delta_y v_{jk} \tag{4.28}$$

Since (4.27) and (4.28) are treated in a similar manner, only the solution for the streamfunction is discussed. Expressing u, v, Ψ' in terms of their Fourier transforms $U_{\ell m}$, $V_{\ell m}$, $\Psi_{\ell m}$ one obtains

$$u_{jk} = \sum_{\ell=0}^{L-1} \sum_{m=0}^{M-1} U_{\ell m}\, e^{\frac{i2\pi j\ell}{L}}\, e^{\frac{i2\pi km}{M}} \tag{4.29}$$

$$v_{jk} = \sum_{\ell=0}^{L-1} \sum_{m=0}^{M-1} V_{\ell m}\, e^{\frac{i2\pi j\ell}{L}}\, e^{\frac{i2\pi km}{M}} \tag{4.30}$$

and

$$\Psi'_{jk} = \sum_{\ell=0}^{L-1} \sum_{m=0}^{M-1} \Psi_{\ell m}\, e^{\frac{i2\pi j\ell}{L}}\, e^{\frac{i2\pi km}{M}} \tag{4.31}$$

Equation (4.27) may then be written explicitly as

$$(\Psi'_{j+1,k} + \Psi'_{j-1,k} - 2\Psi'_{j,k})/(\Delta x)^2 + (\Psi'_{j,k+1} + \Psi'_{j,k-1} - 2\Psi'_{j,k})/(\Delta y)^2 =$$

$$\frac{v_{j,k} - v_{j-1,k}}{\Delta x} - \frac{u_{j,k} - u_{j,k-1}}{\Delta y} \tag{4.32}$$

Upon substitution from (4.29), (4.30), and (4.31) one obtains

$$\left[\sum_{\ell=0}^{L-1} \sum_{m=0}^{M-1} \Psi_{\ell m}\, e^{\frac{i2\pi(j+1)\ell}{L}}\, e^{\frac{i2\pi km}{M}} + \sum_{\ell=0}^{L-1} \sum_{m=0}^{M-1} \Psi_{\ell m}\, e^{\frac{i2\pi(j-1)\ell}{L}}\, e^{\frac{i2\pi km}{M}} \right.$$

$$\left. - 2\sum_{\ell=0}^{L-1} \sum_{m=0}^{M-1} \Psi_{\ell m}\, e^{\frac{i2\pi j\ell}{L}}\, e^{\frac{i2\pi km}{M}} \right]/(\Delta x)$$

$$+ \left[\sum_{\ell=0}^{L-1} \sum_{m=0}^{M-1} \Psi_{\ell m} \, e^{\frac{i2\pi j \ell}{L}} \, e^{\frac{i2\pi(k+1)m}{M}} \right.$$

$$+ \sum_{\ell=0}^{L-1} \sum_{m=0}^{M-1} \Psi_{\ell m} \, e^{\frac{i2\pi j \ell}{L}} \, e^{\frac{i2\pi(k-1)m}{M}}$$

$$\left. - 2 \sum_{\ell=0}^{L-1} \sum_{m=0}^{M-1} \Psi_{\ell m} \, e^{\frac{i2\pi j \ell}{L}} \, e^{\frac{i2\pi km}{M}} \right] / (\Delta y)^2$$

$$= \left[\sum_{\ell=0}^{L-1} \sum_{m=0}^{M-1} v_{\ell m} \, e^{\frac{i2\pi j \ell}{L}} \, e^{\frac{i2\pi km}{M}} \right.$$

$$\left. - \sum_{\ell=0}^{L-1} \sum_{m=0}^{M-1} v_{\ell m} \, e^{\frac{i2\pi(j-1)\ell}{L}} \, e^{\frac{i2\pi km}{M}} \right] / \Delta x$$

$$- \left[\sum_{\ell=0}^{L-1} \sum_{m=0}^{M-1} u_{\ell m} \, e^{\frac{i2\pi j \ell}{L}} \, e^{\frac{i2\pi km}{M}} \right.$$

$$\left. - \sum_{\ell=0}^{L-1} \sum_{m=0}^{M-1} u_{\ell m} \, e^{\frac{i2\pi j \ell}{L}} \, e^{\frac{i2\pi(k-1)m}{M}} \right] / \Delta y \qquad (4.33)$$

Combining terms yields

$$\sum_{\ell=0}^{L-1} \sum_{m=0}^{M-1} \Psi_{\ell m} \, e^{\frac{i2\pi j \ell}{L}} \, e^{\frac{i2\pi km}{M}} \left[\frac{e^{\frac{i2\pi\ell}{L}} + e^{-\frac{i2\pi\ell}{L}} - 2}{\Delta x^2} \right.$$

$$+ \frac{e^{\frac{i2\pi m}{M}} + e^{-\frac{i2\pi m}{M}} - 2}{\Delta y^2} \Bigg]$$

$$= \sum_{\ell=0}^{L-1} \sum_{m=0}^{M-1} v_{\ell m}\, e^{\frac{i2\pi j\ell}{L}}\, e^{\frac{i2\pi km}{M}} \left[\frac{1 - e^{-\frac{i2\pi\ell}{L}}}{\Delta x} \right]$$

$$+ \sum_{\ell=0}^{L-1} \sum_{m=0}^{M-1} u_{\ell m}\, e^{\frac{i2\pi j\ell}{L}}\, e^{\frac{i2\pi km}{M}} \left[\frac{1 - e^{-\frac{i2\pi m}{M}}}{\Delta y} \right] \qquad (4.34)$$

or

$$\sum_{\ell=0}^{L-1} \sum_{m=0}^{M-1} \Psi_{\ell m}\, e^{\frac{i2\pi j\ell}{L}}\, e^{\frac{i2\pi km}{M}} \left[\frac{2\left(\cos\frac{2\pi\ell}{L} - 1\right)}{\Delta x^2} + \frac{2\left(\cos\frac{2\pi m}{M} - 1\right)}{\Delta y^2} \right]$$

$$= \sum_{\ell=0}^{L-1} \sum_{m=0}^{M-1} e^{\frac{i2\pi j\ell}{L}}\, e^{\frac{i2\pi km}{M}} \left[v_{\ell m}\, \frac{1 - e^{-\frac{i2\pi\ell}{L}}}{\Delta x} + u_{\ell m}\, \frac{1 - e^{-\frac{i2\pi m}{M}}}{\Delta y} \right]$$

$$(4.35)$$

The orthogonality condition of the Fourier functions leads to

$$\Psi_{\ell m} = \left[\frac{v_{\ell m}}{\Delta x}\left(1 - e^{-\frac{i2\pi\ell}{L}}\right) + \frac{u_{\ell m}}{\Delta y}\left(1 - e^{-\frac{i2\pi m}{M}}\right) \right] \Bigg/$$

$$\left[\frac{2}{\Delta x^2}\left(\cos\frac{2\pi\ell}{L} - 1\right) + \frac{2}{\Delta y^2}\left(\cos\frac{2\pi m}{M} - 1\right) \right] \qquad (4.36)$$

Ψ' is then computed using Equation (4.31), and finally the grid point value of the streamfunction can be obtained by adding the tilde and prime terms as

$$\Psi_{jk} = \tilde{\Psi}_k + \Psi'_{jk} \tag{4.37}$$

Program (*STREAM*) computes the streamfunction using the relaxation and Fourier methods. The normalized streamfunctions obtained from the two techniques are presented in Figure (4.2). The significant differences between the two methods are found along the boundaries; elsewhere the patterns appear similar.

program *STREAM*

```
c
c     This program computes the streamfunctions from
c     the wind field using the relaxation and the
c     Fourier methods.
c     l           number of points in the east-west direction
c     m           number of points in the north-south direction
c     np          number of pressure levels in the input data set
c     slat        southernmost latitude
c     grid        grid distance
c
      parameter (l=21,m=13,np=7,l1=l-1,m1=m-1,l2=l-2,m2=m-2)
      real    datau (l,m,np),datav(l,m,np),zinv(100)
      real    u(l,m),v(l,m),dx(m),z(100)
      real    psi(l,m),a(l,m),work(2*l)
      complex uu(l,m),vv(l,m)
      open (20,file='uv21.dat',status='old',readonly)
      open (30,file='psirf.dat',status='unknown')
c
c     Read the wind components (1000 to 100 mb) from unit 20.
c
 878      format(10f8.2)
      do 4100 ip = 1, np
          read (20,878) ((datau(i,j,ip),i=1,l),j=1,m)
          read (20,878) ((datav(i,j,ip),i=1,l),j=1,m)
 4100     continue
c
c     Select wind field at 500 mb
c
      do 4102 i = 1, l
      do 4102 j = 1, m
          u   (i,j) = datau (i,j,4)
          v   (i,j) = datav (i,j,4)
```

```
 4102   continue
c
c    Define the grid spacing and the invariant
c    constants for the domain.
c
         slat      = -15.
         grid      = 2.5
         pi        = 4.0*atan(1.0)
         rad       = pi/180.
         dy        = 111.1 * 1000. * grid
         do 4104 j = 1, m
            alat   = (slat + (j-1)*grid)*rad
            dx(j)  = dy * cos(alat)
 4104   continue
         do 4106 j = 1, m
            z(j)   = dx(j)*dx(j)
            zinv(j) = 1./z(j)
 4106   continue
         zz        = dy*dy
         zzinv     = 1./zz
c
c    Define the forcing function (relative vorticity)
c
         do 4108 j = 2, m1
         do 4110 i = 2, l1
 4110       a(i,j)  = (v(i+1,j)-v(i-1,j))/(2.*dx(j))
      &                  -(u(i,j+1)-u(i,j-1))/(2.*dy)
            a(1,j)  = (v(2,j)-v(l1,j))/(2.*dx(j))
      &                  -(u(1,j+1)-u(1,j-1))/(2.*dy)
            a(l,j)  = a(1,j)
 4108   continue
         do 4112 i = 1, l
            a(i,1)  = 2.*a(i,2)-a(i,3)
 4112       a(i,m)  = 2.*a(i,m1)-a(i,m2)
c
c    Compute the net mass out-flux.the outward velocity
c    is corrected to yield a net outward mass flux.
c
c    vno  is the integral mass flux.
c    uno  is the integral of the magnitude of the mass flux.
c
         vno       = v(1,m)*dx(m)/2.+v(l,m)*dx(m)/2.
         uno       = abs(v(1,m))*dx(m)/2.+abs(v(l,m))*dx(m)/2.
```

```
c
      do 4114 i = 2, 11
          uno   = uno + abs (v(i,m))*dx(m)
4114      vno   = vno + v(i,m)*dx(m)
      vno       = vno + u(l,m)*dy/2. + u(l,1)*dy/2.
      uno       = uno + abs(u(l,m))*dy/2.+abs(u(l,1))*dy/2.
c
      do 4116 j = 2, m1
          uno   = uno + abs(u(l,j))*dy
4116      vno   = vno + u(l,j)*dy

      vno       = vno - v(l,1)*dx(1)/2.-v(1,1)*dx(1)/2.
      uno       = uno + abs(v(l,1))*dx(1)/2.+abs(v(1,1))*dx(1)/2.
c
      do 4118 i = 2, 11
          uno   = uno + abs(v(i,1))*dx(1)
4118      vno   = vno - v(i,1)*dx(1)
      vno       = vno - u(1,1)*dy/2.-u(1,m)*dy/2.
      uno       = uno + abs(u(1,1))*dy/2.+abs(u(1,m))*dy/2.
c
c     Computation of the correction factor epsilon
c
      do 4120 j = 2, m1
          uno   = uno + abs(u(1,j))*dy
4120      vno   = vno - u(1,j)*dy
      eps       = vno/uno
      write(6,798)
      write(6,799) uno,vno,eps
  798 format(2x,'uno ,vno , eps')
  799 format(6x,3e14.5)
c
c     Correction of the outward normal velocity.
c
      do 4122 i = 1, 1
          v(i,1) = v(i,1) + eps*abs(v(i,1))
4122      v(i,m) = v(i,m) - eps*abs(v(i,m))
c
      do 4124 j = 1, m
          u(1,j) = u(1,j) + eps*abs(u(1,j))
4124      u(l,j) = u(l,j) - eps*abs(u(l,j))
c
c     Assume psi(1,m) is known and compute the remaining
c     boundary values using the corrected outward normal velocity
```

```
        psi(1,m)   = 0.
        do 4126 i = 2, l
 4126   psi(i,m)    = psi(i-1,m) + (v(i,m)+v(i-1,m))*dx(m)/2.
        do 4128 jj= 1, m1
            j      = m-jj
 4128   psi(l,j)    = psi(l,j+1) + (u(l,j)+u(l,j+1))*dy/2.
        do 4130 ii= 1, l1
            i      = l-ii
 4130   psi(i,1)    = psi(i+1,1) - (v(i,1)+v(i+1,1))*dx(1)/2.
        do 4132 j = 2, m1
 4132   psi(1,j)    = psi(1,j-1) - (u(1,j)+u(1,j-1))*dy/2.
c
c     Solve the Poisson equation using the relaxation technique.
c     The tolerance factor is set to 1000.
c
        call RELAX (psi,zzinv,z,a,zinv,l,l1,m1,m)
c
c     Write output to tape 30.
c
        write (30,222)psi
  222   format(10e13.6)
c
c     Compute the steamfunction via Fourier expansions.
c
        do 4136 j = 1, m
        do 4136 i = 1, l
            uu(i,j) = cmplx (u(i,j),0.0)
            vv(i,j) = cmplx (v(i,j),0.0)
 4136   continue
c
        call PSICHI (uu,vv,l,m,dx,dy,work,1,1)
c
        do 4138 j = 1, m
        do 4138 i = 1, l
            psi(i,j) = real (uu(i,j))
 4138   continue
c
c     Write output to tape 30.
c
        write (30,222)psi
c
        stop
        end
```

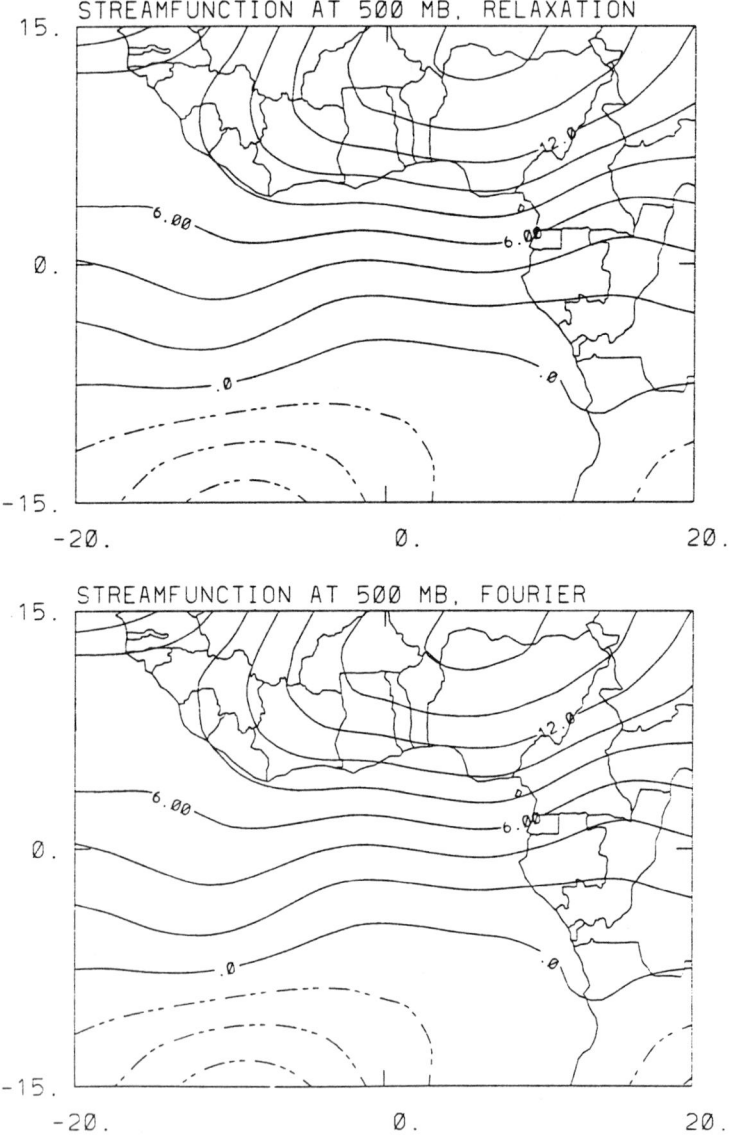

Figure 4.2: Streamfunctions obtained from relaxation and Fourier
methods. Normalization factor $10^{-6}m^2s^{-1}$. (Dashed
values are negative.)

3. Geopotential Height from Wind Field

Various versions of the balance laws are presented in this section and are used to estimate the geopotential height from the wind field. These are obtained from simplifications of the divergence equation which may be written as

$$\nabla^2(gz) = \frac{\partial}{\partial t}(\nabla.\vec{V}) + \vec{V}.\nabla(\nabla.\vec{V}) + \omega\frac{\partial}{\partial p}(\nabla.\vec{V}) + (\nabla.\vec{V})^2 + \nabla.f\nabla\psi$$

$$+ 2J(\frac{\partial\psi}{\partial x},\frac{\partial\psi}{\partial y}) - \frac{\partial\omega}{\partial x}\frac{\partial u}{\partial p} + \frac{\partial\omega}{\partial y}\frac{\partial v}{\partial p} \qquad (4.38)$$

This equation is obtained by simply applying the delta $(\nabla.)$ operator to the equations of motion. The solution of the reverse balance equation provides an estimate for the geopotential field. Three versions of the above balance equation are presented for the calculation of the height field.

i. The geostrophic balance

$$\nabla^2(gz) = f_0\nabla^2\psi \qquad (4.39)$$

ii. The linear balance

$$\nabla^2(gz) = \nabla.f\,\nabla\psi \qquad (4.40)$$

iii. The nonlinear balance

$$\nabla^2(gz) = \nabla.f\nabla\psi + 2J(\frac{\partial\psi}{\partial x},\frac{\partial\psi}{\partial y}) \qquad (4.41)$$

In these three cases the forcing functions are supposedly known and the problem reduces to the solution of Poisson equation,

$$\nabla^2\phi = f(x,y) \qquad (4.42)$$

where $\phi = gz$ is the geopotential. As discussed in previous chapters, the solution of this type of equation requires appropriate specification of the boundary conditions.

In tropical regions it is simpler to avoid the specification of the boundary conditions by assuming periodicity of the observed wind field. Moreover, this boundary condition precludes the need for any assumption in the wind-pressure relationship near the equatorial regions. On the other hand, it is desirable to keep the meridional boundary in the middle latitudes where the geopotential can be obtained either from observations or from the geostrophic relations. Once the boundary conditions are defined, the problem becomes straightforward using the Poisson solver routine. As an illustration, a computer program (*GEOPOTENTIAL*) with different calling options calculating the nonlinear, the linear and the geostrophic reverse balance equations is presented. This driver makes use of subroutine *ZFIELD* which is capable of computing the geopotential field using the three forms of the balance law. The basic principle consists of calculating the streamfunction from the wind field and then the geopotential height from the streamfunction.

Fig. (4.3) shows the geopotential field at 500 mb obtained from the geostrophic balance. The geopotential obtained using the nonlinear and linear balance laws present some similarities (Fig. 4.4).

The obtention of geopotential height from the wind field is a very useful step in the definition of the initial state in numerical weather prediction. It is recognized that these assumptions are just first order approximations and that large departures from these balances may occur in divergent situations. In such situations, the model physics would, in principle, generate divergence and compensate for the departures from the balance laws during the integration of the model.

```
          program GEOPOTENTIAL
c
c     This program computes the geopotential height field
c     from the streamfunction using 3 different methods:
c     l      is the number of points in the east-west direction
c     m      is the number of points in the north-south direction
c     np     is the number of pressure levels in the input data set
c
c     i    :    the geostrophic balance.
c     ii   :    the non-linear balance.
c     iii  :    the linear balance.
c
```

```
          parameter (l=21,m=13,np=7)
          real   dx (m),cor(m),a2(l,m),z500(l,m)
          real   psi(l,m),upsi(l,m),vpsi(l,m)
c
c     Set the mean height at 500 mb and define some constants.
c
          data zbar,beta,slat,grid /5600.,2.0e-11,-15.,2.5 /
c
c     Note that the streamfunction calculated by program
c     STREAM is used as input.
c
          open (30,file='psirf.dat',status='old')
          open (40,file='geop.dat',status='unknown')
c
c     Read streamfunction field from tape 30.
c
 1000    format(10e13.6)
          read(30,1000)psi
c
c     Define the meridional grid spacing and
c     the Coriolis parameter.
c
          pi          = 4.0*atan(1.0)
          rad         = pi/180.
          dy          = 111.1 * 1000. * grid
          do 4200 j = 1, m
             alat     = (slat + (j-1)*grid)*rad
             dx(j)    = dy * cos(alat)
             cor(j)   = 2.*7.29/1e05*sin(alat)
 4200    continue
c
c     Compute geopotential via the geostrophic balance.
c     the flags fk1 and fk2 are set before the call
c
          fk1         = 0.
          fk2         = 0.
          call ZFIELD ( psi,upsi,vpsi,l,m,dx,dy,cor,
     &                  a2, z500, fk1, fk2,zbar,beta )
c
c     Write output.
c
          write (40,1000)z500
c
```

```
c       Reinitialize the work arrays to zero.
c
        call ZERO (upsi,vpsi,l,m)
        call ZERO (a2,z500,l,m)
c
c       Compute geopotential via the linear balance.
c
        fk1      = 1.
        fk2      = 0.
        call ZFIELD (psi,upsi,vpsi,l,m,dx,dy,cor,
     &                   a2, z500, fk1, fk2,zbar,beta)
c
c       Write output.
c
        write (40,1000)z500
c
c       Reinitialize the work arrays to zero.
c
        call ZERO (upsi,vpsi,l,m)
        call ZERO (a2 ,z500,l,m)
c
c       Compute geopotential via the nonlinear balance.
c
        fk1      = 1.
        fk2      = 1.
        call ZFIELD ( psi,upsi,vpsi,l,m,dx,dy,cor,
     &                   a2,z500, fk1, fk2,zbar,beta)
c
c       Write output.
c
        write (40,1000)z500
        stop
        end
```

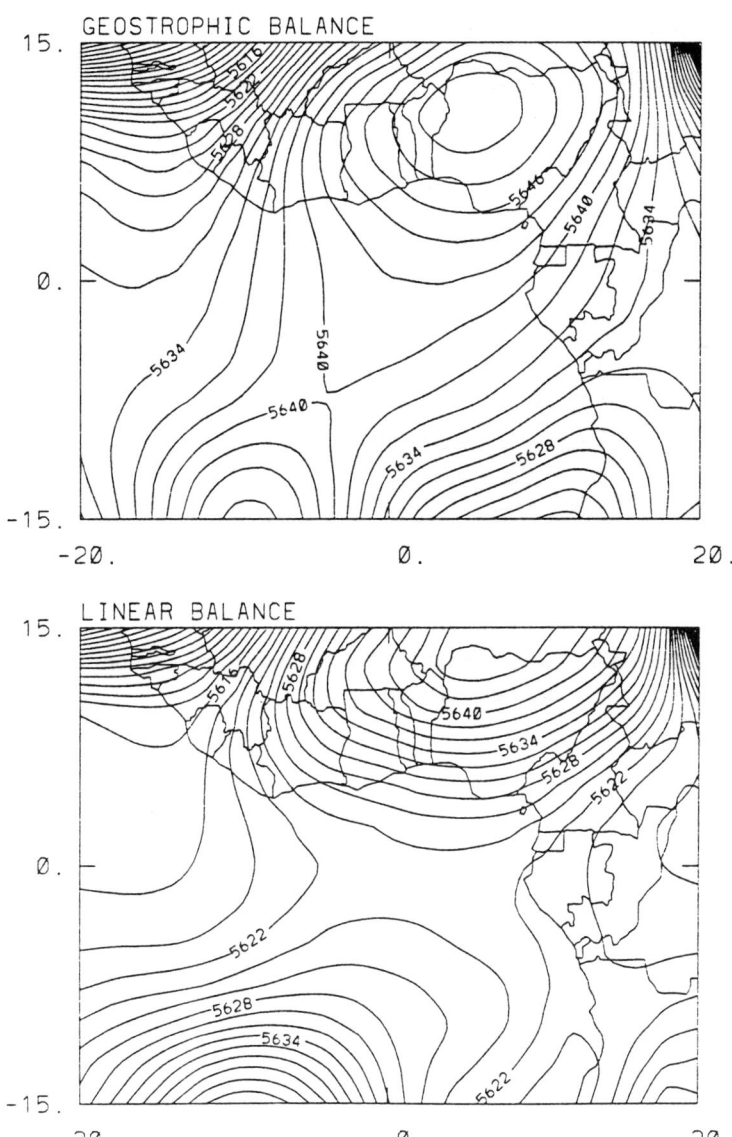

Figure 4.3: Geopotential field at 500 mb from the geostrophic balance and linear balance (meters).

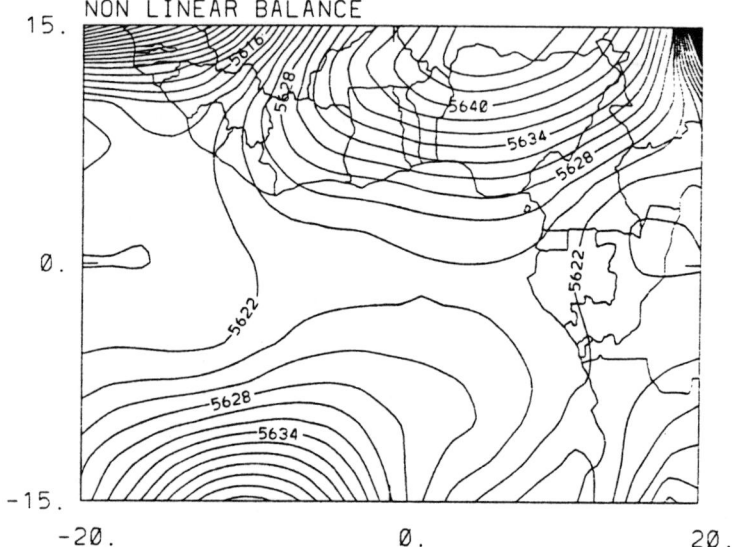

Figure 4.4: Geopotential field at 500 mb from the non-linear
balance (meters).

5

Objective Analysis

Objective analysis is defined as the process which transforms information from randomly spaced observing sites into data at regularly spaced gridpoints (Figure 5.1). Beside its reproducibility, an objective analysis scheme should perform a smooth interpolation, detect and remove bad data and carry out internal consistency analysis.

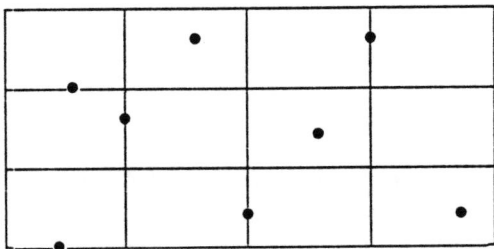

Figure 5.1: Sampling distribution.

In this chapter the theory and the code of three well known methods are presented in a chronological order. Namely, these are Panofsky, Cressman and Barnes' methods which are based on different techniques. An introduction to an optimal interpolation scheme is also described at the end of the chapter.

1. Panofsky's Method, Polynomial Approach

One of the first major contributions to objective analysis was suggested by Panofsky (1949). It consists of fitting a cubic polynomial to a given data set over a region. Panofsky's approach is based on a two dimensional Taylor's expansion,

$$f(x_0 + x, y_0 + y) = \sum_{n=0}^{\infty} \frac{1}{n!} \left[x \frac{\partial}{\partial x} + y \frac{\partial}{\partial y} \right]^n f(x_0, y_0) \tag{5.1}$$

The series can be truncated at the third order to give

$$f(x_0 + x, y_0 + y) = f(x_0, y_0) + x\, f_{0x} + y\, f_{0x} + \frac{x^2}{2!} f_{0xx} + \frac{y^2}{2!} f_{0yy}$$

$$+ xy\, f_{0xy} + \frac{y^3}{3!} f_{0yyy} + \frac{x^3}{3!} f_{0xxx} + \frac{xy^2}{2!} f_{0xyy} + \frac{x^2 y}{2!} f_{0xxy} + R \tag{5.2}$$

where $f_{0x} \equiv \frac{\partial f}{\partial x}(x_0, y_0)$, $f_{0y} \equiv \frac{\partial f}{\partial y}(x_0, y_0)$ and R is the remainder term. In compact form this equation may be written as

$$\tilde{f}(x,y) = \sum_{ij} a_{ij}\, x^i y^j \qquad i + j \leq 3\,;\, i,j \geq 0 \tag{5.3}$$

where, x and y are locations of the known observations $\tilde{f}\,(x,y)$. The solution of this system requires at least 10 independent observations. The use of more observations is, however, believed to enhance error suppression. Writing this equation for all observations yields

$$\begin{bmatrix} 1 & x_1 & y_1 & x_1 y_1 & x_1^2 & y_1^2 & x_1^2 y_1 & x_1 y_1^2 & x_1^3 & y_1^3 \\ 1 & x_2 & y_2 & x_2 y_2 & x_2^2 & y_2^2 & x_2^2 y_2 & x_2 y_2^2 & x_1^3 & y_1^3 \\ & & & & & \cdot \\ & & & & & \cdot \\ & & & & & \cdot \\ 1 & x_n & y_n & x_n y_n & x_n^2 & y_n^2 & x_n^2 y_n & x_n y_n^2 & x_n^3 & y_n^3 \end{bmatrix} \begin{bmatrix} a_{00} \\ a_{10} \\ \cdot \\ \cdot \\ \cdot \\ a_{03} \end{bmatrix} = \begin{bmatrix} f(x_1, y_1) \\ f(x_2, y_2) \\ \cdot \\ \cdot \\ \cdot \\ f(x_n, y_n) \end{bmatrix}$$

$$\tag{5.4}$$

or

$$Aa = z \tag{5.5}$$

where the coefficient matrix is $A(N,10)$, the unknown vector is $a(10, 1)$ and the forcing vector is $z(N,1)$, N being the number of observations. Finally the derivative terms are obtained as

$$a = A^{-1}z \qquad (5.6)$$

In case the matrix is ill conditioned, an iterative scheme will improve the solution. That is, if the solution is in error by δa, then

$$A(a + \delta a) = z + \delta z \qquad (5.7)$$

Subtracting the exact solution from (5.7) leads to

$$A\delta a = A(a + \delta a) - z \qquad (5.8)$$

Since the coefficient matrix is known, (5.8) can be solved for δa, and an improved solution is then obtained by subtracting δa from the first solution. This process can be repeated iteratively. The solution of this type of problem has been described in Chapter 3 and is not repeated here. An output example is however shown for illustration (Figure 5.2).

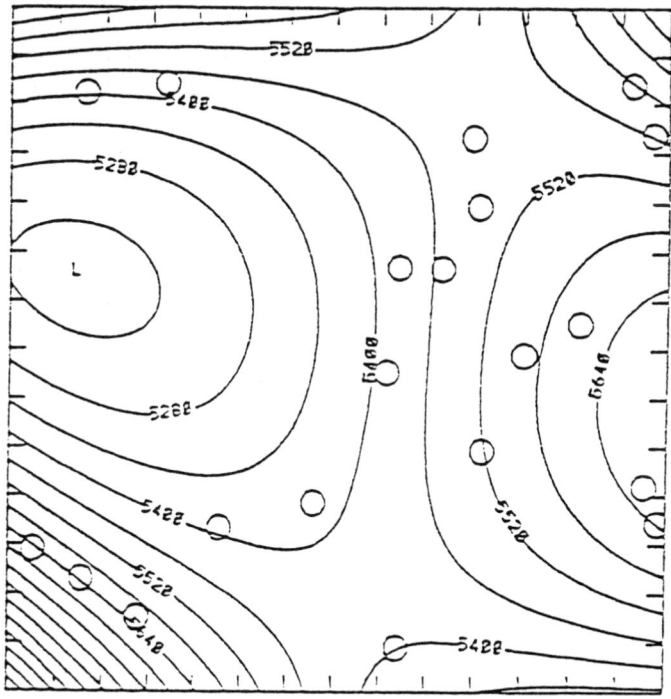

Figure 5.2: Cubic polynomial fitting over a region using the Panofsky's method. Circles indicate station data.

2. Cressman's Method, Successive Corrections Technique

One of the most important problems encountered in the practical
application of any objective analysis technique is related to data
reliability and error suppression. The analysis scheme proposed by
Panosfsky did not prove practical for use over wide regions with poor
data coverage. In addition to the unstabilities developed by the scheme,
the polynomial technique appeared computationally heavy.

The operational numerical weather prediction requires fast, stable
and accurate objective analysis schemes which can be applied to any
meteorological field. The successive corrections method, first introduced
by Bergthorsson and Döös (1955) and developed by Cressman (1959)
offered a first such analysis. The scheme uses a weighted linear sum of
the differences between observations and a guess field at the stations. It
also makes use of a circular area of influence about a given grid-point
(Figure 5.3). At each station, an observation and a guess field are
available. Only an initial guess field is provided at the gridpoints.
Although the choice of the guess field can be arbitrary, to ease the
computational burden it is preferable to have it as accurate an
approximation to the final field as possible. Cressman's technique
involves the successive modification of an initial guess field on the basis
of observed data.

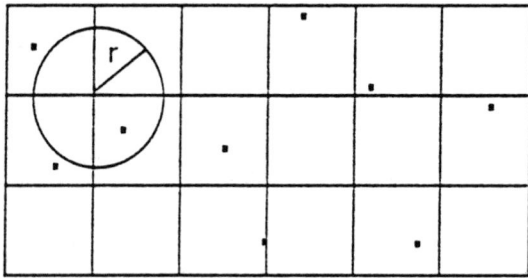

Figure 5.3: The radius of influence and the scanning region.

The development presented in this book follows closely the work of
Tripoli and Krishnamurti (1975). Consider a guess field of some
meteorological variable $X_{g_{ij}}^{(v)}$ on the v-th iteration at gridpoint (i,j). The
guess value at the stations are computed differently depending on
whether the station is located inside or outside the grid box. Within the
grid box the interpolated guess field at a station is obtained as

$$X_{g_s}^{(v)} = \sum_{\substack{i=I-1 \\ j=J-1}}^{\substack{i=I+1 \\ j=J+1}} w_{ij} X_{g_{ij}}^{(v)} \qquad (5.9)$$

where I and J represent the number of grid points in the zonal and meridional directions, respectively. The weighting function, w, is given at each grid point by

$$w_{ij} = \prod_{\substack{k=I-1 \\ k \neq I}}^{k=I+1} \frac{(x - x_k)}{(x_I - x_k)} \prod_{\substack{l=J-1 \\ l \neq J}}^{l=J+1} \frac{(y - y_l)}{(y_J - y_l)} \qquad (5.10)$$

where x and y are grid and data locations, respectively. For stations outside the grid box,

$$X_{g_s}^{(v)} = \frac{\sum_{\substack{i=1 \\ j=1}}^{\substack{i=L \\ j=M}} w_{ij} X_{g_{ij}}^{(v)}}{\sum_{\substack{i=1 \\ j=1}}^{\substack{i=L \\ j=M}} w_{ij}} \qquad (5.11)$$

In this case, the Cressman weighting function is defined as

$$w_{ij} = \begin{cases} \dfrac{R^2 - d^2}{R^2 + d^2} & d < R \\[3mm] 0 & d \geq R \end{cases} \qquad (5.12)$$

where R is the radius of influence and d is the distance between the gridpoint and the station expressed as

$$d = \sqrt{\Delta x^2 + \Delta y^2} \qquad (5.13)$$

The discrepancy between the actual value (X_s) and the interpolated value (X_{gs}) at the station at iteration v is obtained as

$$E_s^{(v)} = X_s^{(v)} - X_{gs}^{(v)} \tag{5.14}$$

The value of $E_s^{(v)}$ is next used to correct the nearby gridpoint value at the $(v + 1)$-th iteration. The correction factor is expressed as

$$c_{ij}^{(v+1)} = \frac{\sum_s W_s^{(v)} E_s^{(v)}}{\sum_s W_s^{(v)}} \tag{5.15}$$

and the weight $W_s^{(v)}$ is given by

$$W_s^{(v)} = w_{ij}^{(v)} \beta \gamma \tag{5.16}$$

where $w_{ij}^{(v)}$ is the Cressman weight defined by (5.12), β is a time weighting function defined as

$$\beta = e^{-\alpha(\Delta t)^2} \tag{5.17}$$

and γ is a reliability factor subjectively chosen to define the weight assigned to each observation type. Following Tripoli and Krishnamurti (1975) the coefficient α is chosen so that $\beta = 0.1$ when the time separation between observations is 48 h. A smoothing factor is introduced and is defined as

$$\bar{s} = \frac{R-d}{C_g} \tag{5.18}$$

where C_g is the center of gravity of the influencing area estimated as

$$C_g = \frac{R}{2} \tag{5.19}$$

The next iterative guess is then given by

$$X_{g_{ij}}^{(v+1)} = X_{g_{ij}}^{(v)} + \bar{s}\, c_{ij}^{(v+1)} \tag{5.20}$$

The objective analysis program presented in this section (*CRESSMAN*) goes one step further and performs a vector field analysis. This program also computes the root mean square errors for the analyzed u,v fields as well as for the vector field. A final smoothing of the analyzed fields is preformed using a nine point smoother defined by

$$\overline{X}_{g_{ij}} = [4X_{ij} + 2(X_{i+1\ j} + X_{i-1\ j} + X_{i\ j+1} + X_{i\ j-1}) + X_{i+1\ j+1} + X_{i-1\ j+1}$$

$$+ X_{i+1\ j-1} + X_{i-1\ j-1}]\, /\ 16 \tag{5.21}$$

Computations of the mean and variance of the corrected vector components are also performed. A scalar analysis can easily be implemented from the vector field analysis by considering one component of the vector only.

program *CRESSMAN*

```
c
c     This program  performs an objective analysis of a vector
c     field using  Cressman's successive corrections technique.
c
c
c     l            : number of grid points in the zonal direction
c     m            : number of grid points in the meridional direction
c     np           : number of pressure levels in the data set
c     pres         : pressure level for which data is analyzed
c     imax         : maximum station data
c     nmaps        : number of map times to be analyzed
c     np           : number of levels for each map time to be analyzed
c     time         : time of analysis
c     rltmn, rlnmn define the southwestern point
c     rltine, rlnine are grid increment in the x and y directions.
c
          parameter (l = 21,m = 13,np = 7)
          real sumx(l,m),sumy(l,m),icount(l,m),store(l,m)
          real a(3,l,m),udev(1500),vdev(1500),rinfo(7,1500)
          real zstd(3),datau(l,m,np),datav(l,m,np)
```

```
              common ug(l,m),vg(l,m)
              equivalence(icount,store)
              real id
 1000    format(' too many vectors read ')
c
c     Define constants
c
              data zstd/1500.,3171.,12353./
              data pres,imax,g,nmaps,lev,time/700.0,1500,9.81,1,1,12./
              data rltmn,rlnmn,rltinc,rlninc /-15.,-25.,2.5,2.5/
c
c     Definitions :
c     unit 7        : first guess field
c
              open(1,file = 'station.dat',status='old')
              open(7,file = 'uv21.dat',status='old')
              open(8,file = 'out.dat',status='unknown')
c
              do 5100 n = 1,nmaps
                 nvect   = 0
                 id      = 6.0
c
c     Read cloud motion vectors
c
  20     read(1,1004,end=30)rlat,rlon,spd,dir
 1004    format(2f12.2,4x,2f12.2)
c
c     Check validity of input data
c
  12     if (dir.gt.360.) go to 20
              if (dir.eq.999.) go to 20
              if (dir.lt.0.   ) go to 20
              if (spd.le.0.   ) go to 20
              if (spd.eq.999.) go to 20
c
c     Arrange data in array rinfo.This array is described
c     in subroutine OBJAN. Note that for this example the
c     wind speed is in (m/s) and is converted to knots before
c     calling the objective analysis scheme
c
              nvect          = nvect+1
              rinfo(7,nvect) = spd*1.94
              rinfo(6,nvect) = dir
```

```
          rinfo(5,nvect) = time
          rinfo(3,nvect) = rlon
          rinfo(4,nvect) = zstd(lev)
          rinfo(2,nvect) = rlat
          rinfo(1,nvect) = id
c
c     Display input data as they are read.
c
          write(6,1004)rlat,rlon,spd*1.94,dir
          if (nvect .le. 1500) goto 20
          print 1000
          stop ' too many vectors read '
    30    continue
          print *,'number of vector read' ,nvect
c
c     Read the guess field
c
   878    format(10f8.2)
          do 5102 ip = 1,np
             read (7,878) ((datau(i,j,ip),i=1,l),j=1,m)
             read (7,878) ((datav(i,j,ip),i=1,l),j=1,m)
  5102    continue
c
c     Extract the wind field at 500 mb.
c
          do 5104 i  = 1, l
          do 5104 j  = 1, m
             ug(i,j)  = datau (i,j,4)
             vg(i,j)  = datav (i,j,4)
  5104    continue
c
c     Write the observed u,v fields.
c
          call BUFF (ug,l*m,8,2)
          call BUFF (vg,l*m,8,2)
c
c     Alter the wind field by setting its v-component
c     to zero. this is to be used as initial guess field.
c
          call ZERO1 (vg,l,m)
          call BUFF (ug,l*m,8,2)
          call BUFF (vg,l*m,8,2)
          print 1001
```

```
1001    format(///,10x,'begin objective analysis ',i2)
c
c       Perform the objective analysis
c
        call OBJAN (rinfo,ug,vg,l,m,nvect,time,lev,sumx
     &                ,sumy,a,icount,udev,vdev,store,rltmn
     &                ,rlnmn,rltinc,rlninc,zstd)
        call BUFF (ug,l*m,8,2)
        call BUFF (vg,l*m,8,2)
5100    continue
        stop
        end
```

The example presented in this section is designed to show the capability of the successive corrections method. It consists of taking 34 station observations and reconstructing a wind field considering its zonal component as a first guess. The station data used for this example have been made up from the observed field. These data are given in Table 5.1, and their locations are illustrated in Figure 5.4 together with the verifying analysis. The first guess and the analyzed fields are shown in Figure 5.5.

3. Barnes' Objective Analysis Scheme

Although Cressman's objective analysis method had an important impact on operational numerical weather prediction, the introduction in 1964 of Barnes' scheme received a considerable interest, especially among the mesoscale research community. Barnes' technique consists of multiple scans using a weighted linear sum of the observations within a defined region of influence around each gridpoint. For the first scan through the data set, the estimated value of the variable at the gridpoint (i,j) is given by

$$U_{ij}^{e1} = \sum_{s=0}^{N} w(d_s, R) U_s^o \tag{5.22}$$

where U_{ij}^{e1} is the estimate at the first scan, N is the number of data

Table 5.1: Station data used to illustrate Cressman's method; latitude and longitude (degrees), wind speed (knots) and wind direction (degrees).

Latitude	Longitude	Wind speed	Wind direction
−14.50	−24.80	10.01	201.50
−10.50	−25.00	7.00	213.69
− 5.00	−24.80	1.78	319.40
7.50	−24.00	6.50	333.43
12.40	−20.00	13.89	74.60
−14.70	−17.50	13.93	231.21
9.50	−17.50	5.51	280.12
−11.00	−15.00	20.16	262.25
0.00	−15.20	9.84	255.00
7.50	−12.50	10.07	219.52
12.50	−12.30	10.59	23.75
12.50	−10.00	12.49	27.76
−14.60	− 7.70	11.47	303.96
− 5.00	− 7.40	10.67	242.96
10.00	− 7.60	7.41	315.00
−10.80	− 5.00	10.32	274.32
4.70	− 5.00	20.74	251.21
−14.20	0.00	13.29	336.80
0.00	0.20	17.73	275.02
10.10	0.00	18.31	257.14
7.50	2.00	18.31	257.14
− 9.00	7.50	5.65	285.95
12.50	7.50	4.87	203.49
−14.50	10.40	3.10	3.58
10.50	10.00	12.07	258.87
−10.10	12.90	6.81	340.02
− 5.00	15.20	8.71	286.82
0.80	15.00	14.57	260.03
−10.20	17.00	5.82	270.00
12.50	20.00	6.75	219.17
−14.30	22.50	18.72	235.98
7.50	22.50	20.33	283.24
−10.00	25.00	14.36	268.45
0.10	25.00	15.60	286.63

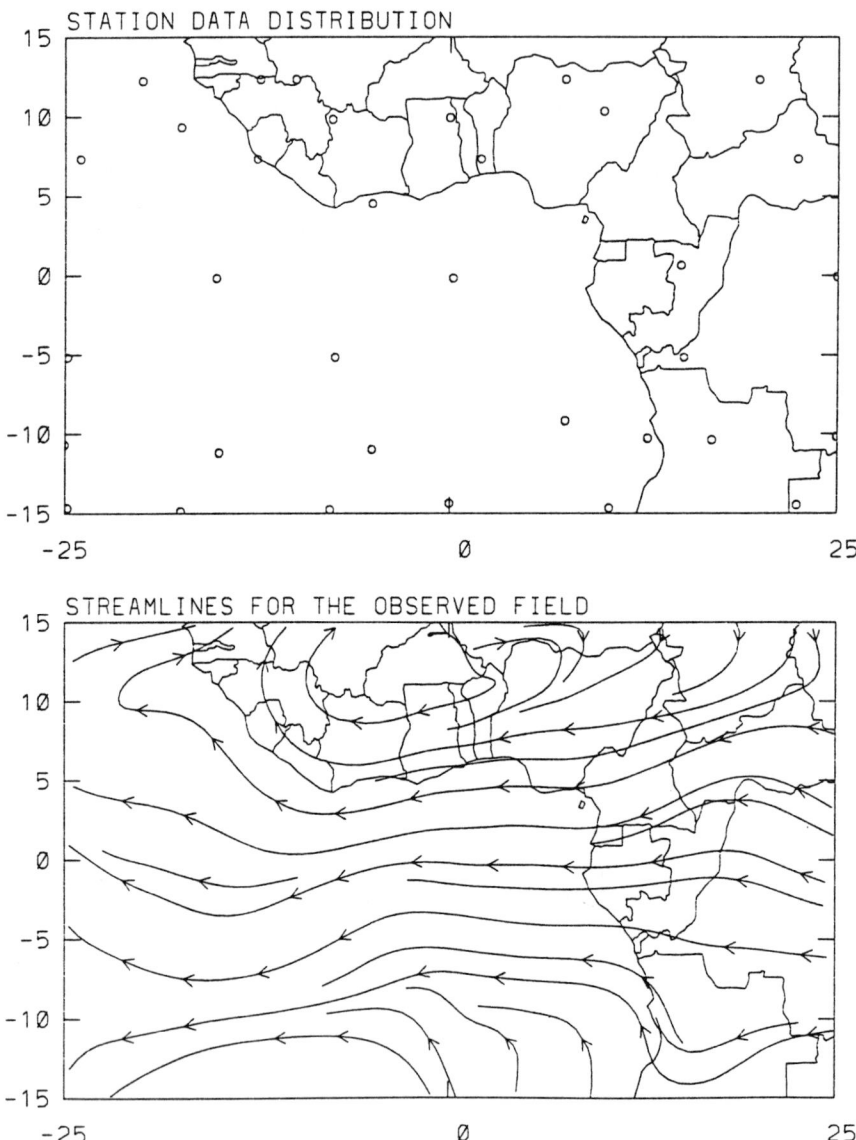

Figure 5.4: Station data distribution and streamlines for the observed
             field.

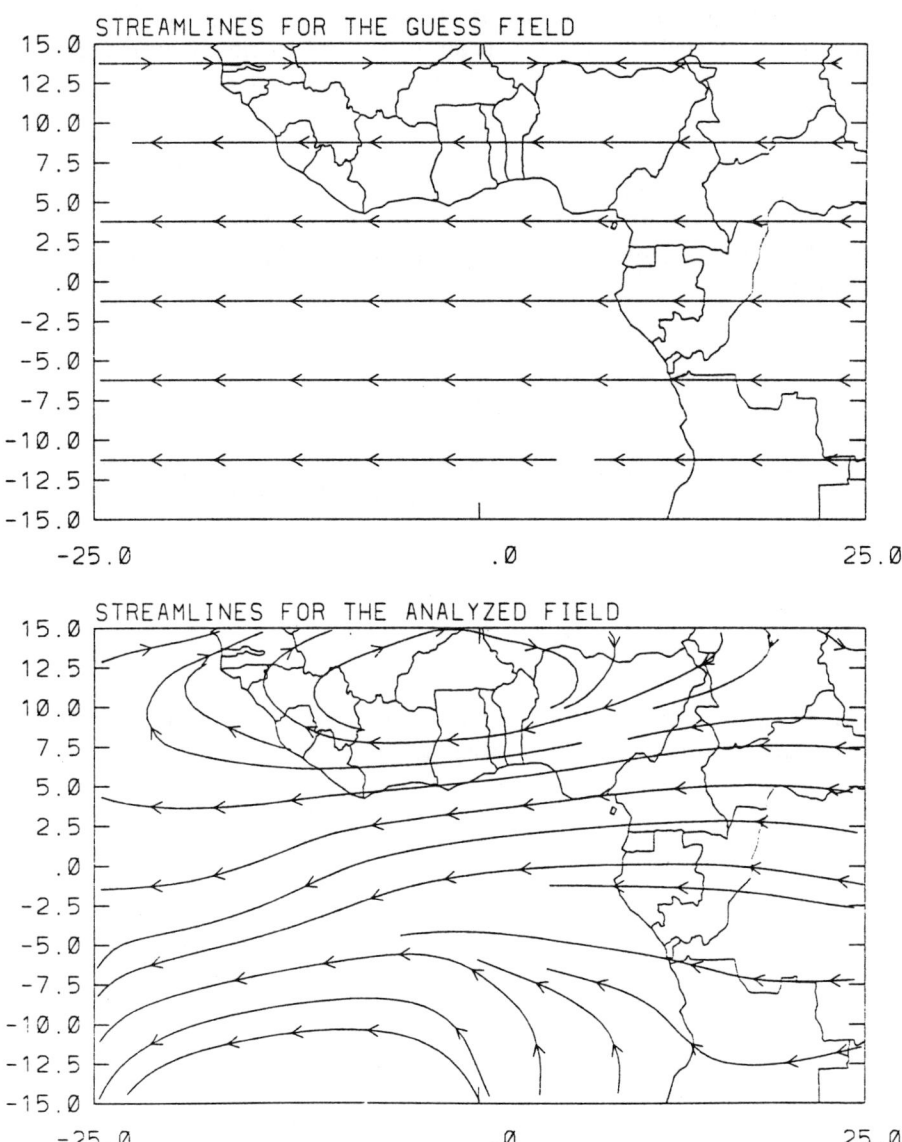

Figure 5.5: Streamlines for the guess field and the analyzed field.

within the region of influence, d_s is the distance between the observation point and the gridpoint, R is the radius of influence and w is the weighting function or filter, which depends on the data density. The first scan appears quite similar to that of Cressman in case the guess field is zero. The estimated field at each station is obtained by simply averaging the four closest gridded values, and the discrepancy at each station is obtained as

$$\Delta^v = U_s^v - U_s^{ev} \tag{5.23}$$

This difference is distributed to all gridpoints using the same weight as for the first scan,

$$U_{ij}^{e(v+1)} = U_{ij}^{ev} + \sum_{s=1}^{N} w(d_s, R)\Delta^v. \tag{5.24}$$

The process is continued until the residual term is less than a prescribed precision factor. The general expression of Barnes' weighting function is defined as

$$w(d) = e^{-\frac{d^2}{4k}} \tag{5.25}$$

k being a parameter defining the shape of the filter's response function. This parameter is obtained by imposing the condition,

$$e^{-\frac{R^2}{4k}} = \varepsilon \tag{5.26}$$

where ε is a small number, chosen such that the weight at d=R is e^{-4} times its maximum value at d=0. This yields

$$k = R^2/16 \tag{5.27}$$

and leads to the ultimate definition of the weighting function as

$$w(d, R) = e^{-\frac{4d^2}{R^2}} \tag{5.28}$$

The interpolated gridpoint field is finally obtained as

$$U_{ij}^e = \frac{\sum\limits_{s=1}^{N} w(d_s)U_s}{\sum\limits_{s=1}^{N} w(d_s)} \tag{5.29}$$

Program **BARNES** presented in this section illustrates an example of scalar analysis using Barnes' method. The example consists of a surface pressure analysis but can be applied to any meteorological field. If a vector field analysis is sought, the vector must be decomposed into its components. It should be noted, however, that no data verification is performed in this example. The user needs then to add some tests in the main program to detect and ignore bad data if suspected. The code is well documented and is relatively straightforward. The station data used for this example are listed in Table 5.2. The analysis output is shown in Figure 5.6.

program *BARNES*

```
c
c      This program performs an objective analysis of a scalar
c      field using Barnes' scheme. It can also be applied to
c      a vector field provided each component is treated
c      separately. The program doesn't include any bad data
c      verification since this depends on the field to be analyzed.
c      A trivial setup has, however, been arranged in subroutine
c      INTERP.
c
       parameter( ns = 85,l = 27,m = 15 )
       common /dist/x(ns),y(ns),ib(ns),jb(ns),xg(l),yg(m)
       real    value(ns),ylat(ns),xlong(ns),hg(l,m)
       data    wlong,slat,dx,dy/97.,39.,0.5,0.5/
c
c      variable definitions.
c
c      ns    : number of station data
c      l     : east-west dimension of the analysis domain
c      m     : north-south dimension of the analysis domain
c      slat  : southernmost latitude
```

```
c      wlong : westernmost longitude
c      dx,dy : grid size in the x and y directions
c
       open (10,file='station.data',status='old     ')
       open (12, file='out6.dat   ',status='unknown')
c
c      Read station data
c      ii     : station number
c      ylat   : station latitude
c      xlong  : station longitude
c      value  : station value of the field to be analyzed
c
  81   format(2x,i2,2x,3f10.2)
       do 5200 i = 1, ns
            read (10,81)ii,ylat(i),xlong(i),value(i)
 5200  continue
c
c      Compute the coordinates of the stations(meters) from the origin
c      (wlong,slat). Also find the coordinates of the closest grid point
c      to the station ,based on the following relationship:
c           iposx = ifact*difx + 1, and
c           iposy = jfact*dify + 1, where
c      ifact and jfact are the number of grid points within 1 degrees of
c      latitude ,and can be expressed as, ifact = integer(1./dx) and
c      jfact = integer(1./dy). Difx,dify are the absolute values of the
c      differences in degrees between the station and the origin,
c      in the x,y directions, respectively.
c
       do 5202 i = 1, ns
            x(i)   = (wlong-xlong(i))*cosd(ylat(i))*1.111e5
            y(i)   = (ylat(i)-slat)*1.111e5
            ib(i)  = 2*(wlong-xlong(i)) + 1
            jb(i)  = 2*(ylat(i) - slat) + 1
 5202  continue
c
c      Compute the distance of all grid points from origin.
c
       do 5204 i = 1, l
            xg(i)  = (i-1)*1.111e5*dx
 5204  continue
       do 5206 j = 1, m
            yg(j)  = (j-1)*1.111e5*dy
 5206  continue
```

```
c
c      Perform the objective analysis.
c
       call OBJAN2 (value,hg,L,M,1)
c
c      Write outputs
c
          write (12,1000) ((hg(i,j),i=1,l),j=1,m)
  1000    format (10f8.2)
          stop
          end
```

Table 5.2: Station data used to illustrate Barnes' method. Latitude and
            longitude (degrees), surface pressure (mb). Missing data are
            represented by 9999.90.

Station	Latitude	Longitude	Surface Pressure
1	45.87	95.38	1014.10
2	45.55	94.07	1014.50
3	44.92	97.00	1014.50
4	44.50	95.08	1014.50
5	44.88	93.22	1014.80
6	44.22	93.92	1017.20
7	43.92	92.50	1015.10
8	43.57	96.73	1014.80
9	43.17	95.15	1015.50
10	42.40	96.38	1014.80
11	43.15	93.33	1014.80
12	42.55	94.18	1016.10
13	42.55	92.40	1014.10
14	41.30	95.90	1014.50
15	40.85	96.75	1014.50
16	41.53	93.65	1013.40
17	42.40	90.70	1012.40
18	41.88	91.70	1011.10
19	41.10	92.45	1012.10
20	40.62	93.93	1013.80
21	40.10	92.55	1013.10
22	40.25	93.72	1015.50

Station	Latitude	Longitude	Surface Pressure
23	39.05	96.77	1013.40
24	39.07	95.63	1013.20
25	39.10	94.60	1012.80
26	39.77	95.92	9999.90
27	39.00	92.22	1012.80
28	39.00	90.38	1013.10
29	44.87	91.48	1015.10
30	45.63	89.45	1015.10
31	44.92	89.62	1015.50
32	44.78	89.67	1015.50
33	44.48	88.13	1015.50
34	44.00	88.57	1014.80
35	44.13	87.67	9999.90
36	44.25	88.52	1015.30
37	43.93	90.27	9999.90
38	43.87	91.25	1014.80
39	43.20	90.18	1014.10
40	43.13	89.33	1013.80
41	42.62	89.03	1013.40
42	43.12	88.05	9999.90
43	42.95	87.91	1013.80
44	42.20	89.10	1013.10
45	41.98	87.90	1013.80
46	42.08	87.82	1012.80
47	41.87	87.62	1013.40
48	41.37	88.68	1012.10
49	41.45	90.52	1010.70
50	40.78	91.12	1011.40
51	39.93	91.20	1012.80
52	40.67	89.68	1011.70
53	40.60	88.90	1012.80
54	39.85	89.67	1012.10
55	40.03	88.28	1012.80
56	40.20	87.62	1012.80
57	45.57	84.80	1015.80
58	44.73	85.58	1016.50
59	44.37	84.68	1016.80
60	43.17	86.25	1015.80
61	45.82	88.12	1017.20
62	45.12	87.63	9999.90
63	45.73	87.08	1015.50

Station	Latitude	Longitude	Surface Pressure
64	43.53	84.08	1016.50
65	42.88	85.52	1016.10
66	42.78	84.60	1017.50
67	42.27	84.47	1017.50
68	42.23	85.57	1016.50
69	42.30	85.23	1016.80
70	42.13	86.43	9999.90
71	41.70	86.32	1014.50
72	41.00	85.20	1015.80
73	40.23	85.38	1015.50
74	39.73	86.28	1014.80
75	40.42	86.93	1013.80
76	39.45	87.30	1013.80
77	39.13	86.82	1013.80
78	41.60	84.00	1016.10
79	39.90	84.22	1016.10
80	39.05	84.67	1015.80
81	43.75	87.69	1015.60
82	42.70	87.10	1013.40
83	43.30	86.40	1015.70
84	42.58	87.82	9999.90
85	44.13	87.55	9999.90

4. Optimum Interpolation Technique

This section discusses a multivariate optimum interpolation procedure currently in use at the Florida State University (FSU). The technique is based on the work conducted at the National Meteorological Center (NMC) (Bergman, 1979; Dey and Morone, 1985; DiMego, 1988). The description of the scheme follows a presentation by Beven (1994).

Many of the objective analysis schemes that have been devised to this date employ methods that have been described as "quasi-statistical". Optimum interpolation techniques, which fall under optimal statistical objective analysis grouping (Thiebaux and Pedder, 1987) use a firm statistical framework as a basis for objective analysis.

Like other objective analysis systems, optimum interpolation minimizes the differences between the observations and the first guess field, and uses weighted observations to estimate the gridpoint values. However, it is different in that its weighting functions are based upon

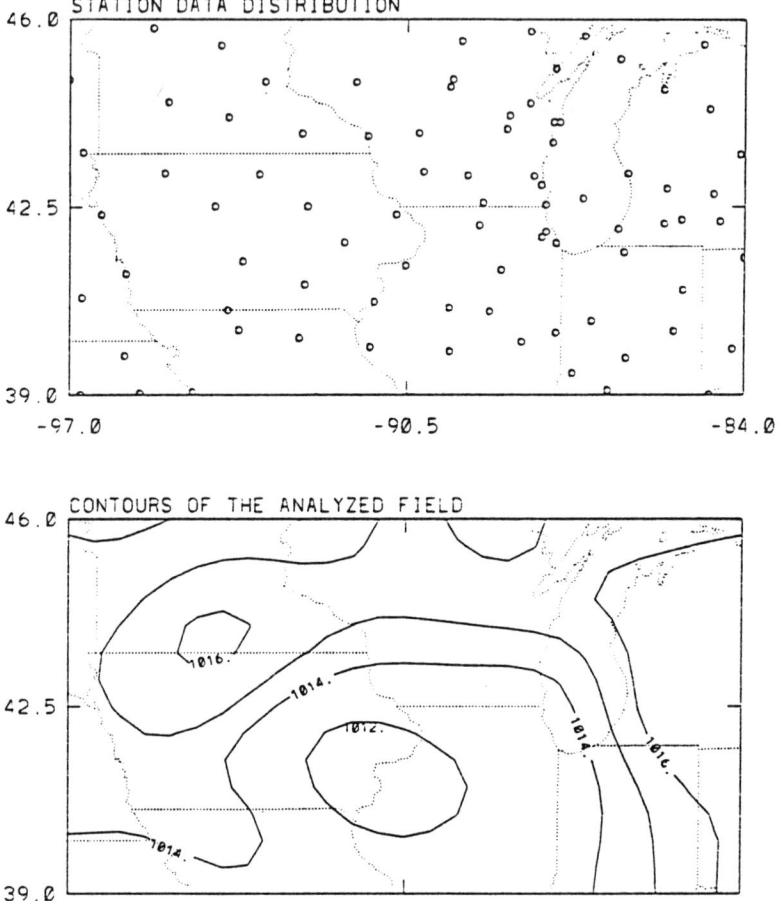

Figure 5.6: Station data distribution and contours of the analyzed field.

statistical analyses of discrepancies between observations and guess fields
as well as errors inherent to observations themselves. This is different
from the empirically determined weighting functions used by Cressman
(1959) and Barnes (1964).

4.1 Formulation of the Optimum Interpolation System

The basic scheme used in the FSU system was developed for NMC global analysis by Bergman (1979), and most of the following discussion will be based on that work. Some modifications were done during the 1980's, including the use of isobaric coordinates and improving the analysis resolution (Dey and Morone, 1985). DiMego (1988) adapted the system for use with the NMC Regional Analysis and Forecast System (RAFS) built around the Nested Grid Model (NGM), using a sigma coordinate system to match the NGM. The FSU system has elements of all of this work, as it is a regional optimum interpolation scheme based on isobaric coordinates.

4.1.1 Basic Equations

As shown by Bergman (1979), the guess field value at an observation site may be determined from bilinear horizontal interpolation and linear vertical interpolation. This results in the following relation between the observed value and the corresponding guess value,

$$\hat{f}_{ik} = \hat{F}_{ik} - \tilde{F}_{ik} \qquad (5.30)$$

where $\hat{f}_k$ is the difference for the ith observation of the kth variable, $\hat{F}_{ik}$ is the observed value, and $\tilde{F}_{ik}$ is the corresponding guess value. According to Bergman (1979), if more than one variable is measured by an instrument at a particular location, each measured quantity counts as a separate observation. Furthermore, each observation carries a measurement error such that

$$\hat{F}_{ik} = F_{ik} + e_{ik} \qquad (5.31)$$

where F_{ik} is the "true" value of the kth variable at the location of the ith observation, and e_{ik} is the measurement error. Similarly, the residual value may be expressed as

$$\hat{f}_{ik} = f_{ik} + e_{ik} \qquad (5.32)$$

where

$$f_{ik} \equiv F_{ik} - \tilde{F}_{ik} \qquad (5.33)$$

is the difference between the true value of the kth variable and the first guess at the ith location. From this, the general form of the analysis approximation may be written as

$$F_{gr} \approx \tilde{F}_{gr} + \sum_{k=1}^{m} \sum_{i=1}^{n_k} a_{ik} \, \hat{f}_{ik} \tag{5.34}$$

where $\tilde{F}_{gr}$ is the guess value of F_{gr} at the gridpoint, m is the number of variables entering the multivariate analysis of variable r at gridpoint g, and n_k is the number of observations of variable k used in the analysis. The term a_{ik} is the weight of the ith observation of type k forming the estimate of F_{gr} (Bergman, 1979). The mean-square error is then obtained as

$$E_{gr}^2 = \overline{[F_{gr} - \tilde{F}_{gr} - \sum_{k=1}^{m} \sum_{i=1}^{n_k} a_{ik}(f_{ik} + e_{ik})]^2} \tag{5.35}$$

where the overbar represents an ensemble average and leads to the following system,

$$\sum_{l=1}^{m} \sum_{j=1}^{n_l} (\overline{f_{ik} \, f_{jl}} + \overline{f_{ik} \, e_{jl}} + \overline{e_{ik} \, f_{jl}} + \overline{e_{ik} \, e_{jl}})a_{jl} = \overline{f_{ik} \, f_{gr}} + \overline{f_{ik} \, f_{gr}}$$

$$k = 1, 2, \ldots, m \qquad\qquad i = 1, 2, \ldots, n_k \tag{5.36}$$

The overbar terms in Equation (5.36) represent covariances, while a_{ik} represents the unknown observation weights. Dividing Equation (5.36) by

$(\overline{f_{ik}^2 \, f_{gr}^2})^{1/2}$ yields the normalized form of the system

$$\sum_{l=1}^{m} \sum_{i=1}^{n_l} (\rho_{ij}^{kl} + \tau_{ij}^{kl}\varepsilon_{jl} + \tau_{jl}^{lk}\varepsilon_{ik} + \eta_{ij}^{kl}\varepsilon_{ik}\varepsilon_{jl})a_{jl} = \rho_{ig}^{kr} + \tau_{gi}^{rk}\varepsilon_{ik}$$

$$k = 1, 2, \ldots, m \qquad\qquad i = 1, 2, \ldots, n_k \tag{5.37}$$

In this equation,

$$\rho_{ij}^{kl} \equiv \overline{(f_{ik} \, f_{jl})} \, / \, (\overline{f_{ik}^2 \, f_{jl}^2})^{1/2} \tag{5.38a}$$

$$\overline{\tau_{ij}^{kl}} \equiv (\overline{f_{ik}\ e_{jl}}) / (\overline{f_{ik}^{2}}\ \overline{e_{jl}^{2}})^{1/2} \tag{5.38b}$$

$$\overline{\eta_{ij}^{kl}} \equiv (\overline{e_{ik}\ e_{jl}}) / (\overline{e_{ik}^{2}}\ \overline{e_{jl}^{2}})^{1/2} \tag{5.38c}$$

$$\varepsilon_{ik} \equiv (\overline{e_{ik}^{2}} / \overline{f_{ik}^{2}})^{1/2} \tag{5.38d}$$

$$a_{ij} \equiv (\overline{f_{jl}^{2}} / \overline{f_{gr}^{2}})^{1/2}\ a_{jl} \tag{5.38e}$$

Equation (5.38a) is a correlation relating the error in the guess value at one location to the corresponding error at another location. Equation (5.38b) is a similar correlation relating the error in the guess value at one location to the observational error at another location. Equation (5.38c) is the correlation relating the observational errors in one location to those in another location. The first and third of these correlations will be discussed in the next section. The second one is assumed to be negligible (Bergman, 1979). Therefore, (5.37) reduces to

$$\sum_{l=1}^{m} \sum_{j=1}^{n_l} (\rho_{ij}^{kl} + \eta_{ij}^{kl}\varepsilon_{ik}\varepsilon_{jl})a_{jl} = \rho_{ig}^{kr}$$

$$k = 1,2, ..., m \qquad\qquad i = 1,2, ... , n_k \tag{5.39}$$

Several things need to be said about this formulation. First of all, much of the analysis is multivariate. In the Bergman (1979) scheme, for example, temperatures, winds, and geopotential heights can contribute to the height analysis at a gridpoint. On the other hand, in the DiMego (1988) scheme, the temperature analysis was done separately and was not allowed to contribute to the multivariate analysis of heights and winds. The FSU system currently allows temperatures to contribute information to height and wind analyses if there are no height and wind observation at that level nearby and the temperature observations are at some level other than that of the considered gridpoint (Xue, 1990). It should be noted that moisture variable analysis is done separately (Bergman, 1979; Dey and Morone, 1985; DiMego, 1988). Finally, the system of equations defined by (5.37) gives the solution for one gridpoint.

The Bergman (1979) scheme was tested with differing numbers of pieces of information that were allowed into the analysis for one gridpoint. The version of the FSU system presented here is limited to 24 observations due to computational limitations. Generally, it is desirable to have many observations influence the analysis, which results in a smooth analysis. It appears that too few observations can make for a rough-looking analysis.

The guess value errors need to be determined by careful examination of the errors in the system that produce the first guess. Bergman (1979) suggests that these values vary with height for wind and height fields. Values employed by Bergman (1979) are given in Table 5.3. These values were derived by comparing the NMC global model 6-h forecasts with rawinsonde observations. Values shown by DiMego (1988) were derived by adding an error growth rate to the analysis error standard deviations. The FSU system currently uses the values given in Table 5.4 for height and wind. The temperature error is set to a constant for all levels, while the relative humidity error is 10% for all levels.

Table 5.3: Mean forecast error standard deviation as a function of pressure. (From Bergman, 1979)

Pressure (mb)	T ($^\circ$C)	U (ms^{-1})	V (ms^{-1})
1000	4.3	4.9	4.6
850	3.2	4.2	4.1
700	2.2	4.3	4.0
500	2.0	4.9	4.7
400	2.1	5.7	5.6
300	2.6	7.1	6.7
250	3.4	7.5	7.2
200	3.6	7.3	7.0
150	3.1	6.7	6.4
100	3.0	5.6	5.4
70	2.9	6.0	6.3
50	4.6	9.0	9.8

Table 5.4: First guess errors employed in the FSU optimum interpolation system.

Pressure (mb)	Z (m)	U (ms^{-1})	V (ms^{-1})
1000	9.6	1.0	1.0
850	9.6	1.3	1.3
700	9.6	2.2	2.2
500	12.7	2.9	2.9
400	14.8	3.8	3.8
300	17.8	5.8	5.8
250	18.4	5.0	5.0
200	17.9	5.3	5.3
150	17.8	5.5	5.5
100	16.6	5.0	5.0

The observation errors are dependent on the observation system and can vary widely from one to another (Bergman, 1979). Currently, since the FSU system is employing only one instrument type, only one set of errors is needed for each variable. In an interesting aside to these errors, once general values of the observation error are established for a given measuring device, it may be possible to manipulate them to add more information to the analysis. For example, the problem currently being worked on with the FSU system involves many measurements that are not at the same time as the final analysis. By changing the allowed observation error as a function of time, it may be possible to include off-time observations in the optimum interpolation framework without having to compute their correlations. This method is still being tested.

4.1.2 Correlations

Inspection of (5.37) shows the role of correlations in determining the weights. Use of the correlations comes out of the statistical theory on which optimum interpolation is based. The theory is complex and is beyond the scope of this workbook. However, it can be mentioned that it involves joint probability distributions of several stochastic variables and the use of the multivariate normal distribution. A complete description of these concepts can be found in Thiebaux and Pedder (1987). Correlations describe the statistical structure of the atmosphere (Gandin, 1963). Due to this, they change considerably from point to point in the atmosphere depending on how the first point is related to the second. As a result, any set of weights based on correlations will change from point to point. This is a considerable difference from the fixed weights of Cressman and Barnes methods. Since correlations can change drastically, it is important to use the best correlations available.

Much work has gone into the proper choice of correlation functions for optimum interpolation. One simplifying assumption is that the correlations are assumed to be isotropic and homogeneous, which makes them depend only on distance (Thiebaux and Pedder, 1987). While the real atmosphere is anything but homogeneous and isotropic, it is assumed that the deviations from the first guess field are homogeneous and isotropic.

Bergman (1979) used three-dimensional correlations in the optimum interpolation system to relate errors in the guess value to each other (5.38a). This practice has been followed by both the RAFS and FSU systems. The form for these is

$$\rho_{ij}^{ut} = \mu_{ij}^{ut} \, v_{ij}^{ut} \tag{5.40}$$

where μ is the horizontal component and v is the vertical component. The form for the horizontal component may be written as

$$\mu_{ij}^{ut} = \exp[-k_h \, (S_{ij})^2] \tag{5.41}$$

where k_h is a constant with units length squared and S is the distance between the points in question. Bergman (1979) set $k_h = 0.98 \times 10^{-6}$ km^{-2}, although this was varied in some of his experiments. DiMego (1988) suggested that k_h varies with height between 1.5×10^{-6} km^{-2} and 4.125×10^{-6} km^{-2} and that it should also depend on the scale of the analysis. The FSU system currently uses $k_h = 2.0 \times 10^{-6}$ km^{-2} for all levels of the analysis. The vertical component is the same for Bergman (1979), DiMego (1988), and the FSU system and can be expressed as

$$v_{ij}^{ut} = \frac{1}{1 + k_p \, \ln^2(p_i \, / \, p_j)} \tag{5.42}$$

where k_p is a constant and p is the pressure level. As with k_h, there is no general agreement as to where k_p should be set. Bergman (1979) set k_p

= 5. DiMego (1988) uses a k_p that is level dependent and ranged in value from 4 to 18. In the FSU system, currently $k_p = 5$.

Not only are the height (h), the zonal and meridional winds (u, v) and temperature (t) correlated with themselves, but they are also cross-correlated with each other. This introduces a dynamic constraint on the analysis. Many cross-correlations in Bergman (1979) and DiMego (1988) involve a coefficient of geostrophy, which is a measure of how geostrophic the corrections to the guess field are. This coefficient varies with latitude, as it decreases to zero in the tropics and increases to larger values in the high latitudes. The formula given in Bergman (1979) is

$$G = V / V_g = 1 - \exp \left(-0.5 \left| \text{latitude} \right| \right) \qquad (5.43)$$

where V_g is the geostrophic wind. Use of this coefficient is demonstrated by this height-zonal wind cross-correlation function from DiMego (1988)

$$\mu_{h_i} \mu_{u_j} = - \left[\frac{|f_i|}{f_j} \left(2 k_h R^2 \right)^{1/2} \left(\phi_i - \phi_j \right) \right] G_{gp} \, \mu_{h_i} \mu_{h_j} \qquad (5.44)$$

where R is the earth's radius, ϕ is the latitude and f is the Coriolis parameter.

In the optimum interpolation algorithm, not every point is modified by every observation. Bergman (1979) and the FSU scheme use a scan radius of influence around the gridpoint. This can cause an edge effect in the analysis if the correlation values are too high at the scan radius.

The third set of correlations (5.38c) is that of the observational error at one point with the observational error at another. Bergman (1979) mentions that most of the time these errors are assumed to be un-correlated for one reason or another (i.e., different types of observing platforms with different error characteristics). However, there are circumstances mentioned where this might not be appropriate including when devices like rawinsondes are considered. This could also apply to the dropwindsondes. At the present time, the FSU system assumes these correlations are zero.

4.2 Gridded Data

Current practice for the NMC optimum interpolation scheme is to use a first guess field derived from the 6 hr model run of the Global Data Assimilation System (GDAS). The FSU scheme is somewhat different as it uses the European Center for Medium Range Weather Forecasting (ECMWF) analysis as its first guess field. These data are originally on a 2.5 × 2.5° latitude-longitude grid mesh. This is then interpolated to a 99 × 99 grid array with 33 km grid-spacing. These data are for 10 isobaric levels and five variables, which differs from RAFS sigma levels (DiMego, 1988), but is similar to the global optimum interpolation scheme discussed by Dey and Morone (1985).

6

Basic Physical Concepts

The description of the physical processes is a very important part in numerical weather prediction models. However, because the physics governing these processes is complex and not well known, a detailed explicit prediction of their effects appears either impossible or impractical. In the present generation of numerical weather prediction models, only approximate solutions are sought to describe the effects of the different physical mechanisms acting on the earth atmosphere systems. The parameterization of these processes may then differ from one model to another, depending on the numerical simulation's objectives. While these parameterizations may carry different levels of details, they consist generally of common basic concepts.

Important topics related to physical parameterizations in numerical weather prediction models are introduced in this chapter. Basically, the use of moisture variables in meteorological equations is discussed and algorithms describing their computational aspects are presented.

Although moisture variables may be presented in many different forms, the most commonly used in weather prediction are the temperature (T), dewpoint temperature (T_d), the specific humidity (q) and the relative humidity (R_h). Dewpoint depression (T-T_d), mixing ratio and vapor pressure also constitute a useful set of variables. Furthermore, it is often necessary, during the integration of the model, to perform forward and backward conversions between moisture variables. Towards this end, using Teten's formula for the vapor pressure,

$$e_s = 6.11 \exp \frac{a(T - 273.16)}{(T - b)} \tag{6.1}$$

and the saturation specific humidity

$$q_s = 0.622 \frac{e_s}{P - 0.378 \, e_s} \tag{6.2}$$

a set of short routines is developed to obtain any moisture variable provided some others are known. Here, the constants a and b take different values based on whether the saturation occurs over water or ice.

a = 17.26	b = 35.86	over water
a = 21.87	b = 7.66	over ice

In practice, the separating temperature between the two phases is usually taken as $263°K$. The vapor pressure is expressed in mb and the specific humidity in gm/gm. Some useful pairs of moisture variable transforms subroutines are presented in the following section.

1. Conversion of Moisture Variables

1.1 Dewpoint Temperature and Relative Humidity

The problem consists of estimating the dewpoint temperature given the air temperature and relative humidity at different pressure levels, or calculating the relative humidity given the dewpoint and temperature profiles. The calculation of the dewpoint temperature from the relative humidity is done in two steps. The first step consists of computing the saturation specific humidity, $q_s(T)$, at the given air temperature using (6.2) and the specific humidity as $q(T) = q_s(T)*rh/100$. Assuming $T_d = T$ as a first guess, the second step consists of calculating the saturation specific humidity in an iterative way. At each iteration T_d is decreased by an increment of 0.05, and a new $q_s(T_d)$ is calculated using (6.2) The dewpoint temperature is reached when the newly computed saturation specific humidity equals the specific humidity at the air temperature computed in step one.

Inversely, if the relative humidity is sought from a known dewpoint and temperature profile, the specific humidity would be computed from the dewpoint temperature and the saturation specific humidity from the air temperature. The relative humidity is obtained as the ratio of these two quantities and is expressed in percent. Subroutines *TDFRMRH* and *RHFRMTD* are provided for these calculations.

1.2 Specific and Relative Humidity

The computation of one of these two variables when the other is known at given temperature and pressure levels is a trivial exercise using (6.1) and (6.2). Nevertheless, for completeness two routines named *QFRMRH* and *RHFRMQ* are presented to perform these conversions.

1.3 Dewpoint Temperature and Specific Humidity

The problem here consists of cooling the air parcel until it becomes saturated. As in Section (1.1) T_d is set equal to T as a first guess and is next decreased iteratively by 0.05. The saturation specific humidity is calculated at each iteration and is checked against the known specific humidity. The dewpoint temperature is obtained when these two quantities are equal. The calculation of the specific humidity from the dewpoint temperature is straightforward using (6.2) in which T_d is substituted to T. Subroutines *TDFRMQ* and *QFRMTD* perform these conversions.

Program (*CONVERT*), presented in this section, uses sample profiles of temperature and specific humidity to perform the different conversions. Outputs from each of the above mentioned routines are shown in Table 6.1.

```
        program CONVERT
c
c    This programs performs the different conversions
c    between moisture variables. The input data required
c    for this code are supplied through data statements.
c
        parameter (nkt = 46,nk = 39)
        real p(nkt),t(nkt), q(nkt)
        real td(nkt),rh(nkt)
c
c    Sample data used for the different computations.
c
c    Pressure:
c
        data p/ 1011.,1000.,955.4,950.0,900.,850.,816.2,800.,782.4.750.,
     &    722.4,700.,687.1,660.5,655.1,650.,641.2,600.,594.3,578.1,566.3,
     &    550.,544.6,534.9,528.2,517.7,508.9,500.,493.4,477.9,468.7,450.,
     &    444.2,400.,390.3,382.5,369.5,361.5,350.,310.6,300.0,258.2,250.,
     &    235.6,222.5,200./
c
c    Temperature:
c
        data t/23.0,22.7,22.0,21.8,18.4,13.2,15.6,12.4.11.3,9.9.8.8,7.4,
     &      6.0,5.9,5.7,5.3,4.6, 1.1,.6,-.6,-1.2,-1.9,-2.5,-3.9,-4.7,
     &      -4.7,-5.7,-6.0,-6.4,-8.5,-9.6,-11.4,-12.1,-17.9,-18.5,-19.5,
```

```
      &          -21.6,-22.0,-24.0,-31.0,-32.8,-41.4,-43.1,-46.4,-49.9.-56.3/
c
c     Specific humidity:
c
          data q/.0170,.0164,.0155,.0155,.0143,.0116,.0119,.0112..0106,
      &          .0101,.0097,.0091,.0080,.0068,.0066,.0064,.0062..0055,
      &          .0054,.0057,.0051,.0022,.0020,.0020,.0028,.0025..0020,
      &          .0007,.0008,.0012,.0027,.0017,.0015,.0013,.0006..0005,
      &          .0006,.0004,.0003,.0001,.0002,.0001,.0001,.0001..0001,
      &          .0001/
c
c     Compute td from q.
c
          write  (6,1000)
          write  (6,1001)
          do 6100 k = 1, nk
              tt       = t(k) + 273.16
              pp       = p(k)
              qq       = q(k)
              call TDFRMQ (pp,tt,qq,ttd)
              td(k)    = ttd
              write (6,1002)k,pp,tt-273.16,qq,ttd-273.16
 6100     continue
 1000     format (/,10x,'computation of td from q',/)
 1001     format (3x,'lev',3x,'pressure',1x,'tempera.',3x,'sp hum',4x,
      &     'dew pt',/)
 1002     format (4x,i2,4x,f6.1,4x,f5.1,4x,f6.4,4x,f5.1)
c
c     Compute q from td.
c
          write  (6,1003)
          write  (6,1004)
          do 6102 k = 1, nk
              tt       = t(k) + 273.16
              pp       = p(k)
              ttd      = td(k)
              call QFRMTD (pp,tt,ttd,q1)
              write (6,1005)k,pp,tt-273.16,ttd-273.16,q1
 6102     continue
 1003     format (/,10x,'computation of q from td',/)
 1004     format (3x,'lev',3x,'pressure',1x,'tempera.',3x,'dew pt',4x,
      &     'sp hum',/)
 1005     format (4x,i2,4x,f6.1,4x,f5.1,4x,f5.1,4x,f6.4)
```

```
c
c     Compute rh from td.
c
      write  (6,1008)
      write  (6,1006)
      do 6104 k = 1, nk
          tt      = t(k) + 273.16
          pp      = p(k)
          ttd     = td(k)
          call RHFRMTD (pp,tt,ttd,rrh)
          rh(k)   = rrh
          write (6,1007)k,pp,tt-273.16,ttd-273.16,rrh
 6104 continue
 1006 format (3x,'lev',3x,'pressure',1x,'tempera.',3x,'dew pt',4x,
     &  're hum',/)
 1007 format (4x,i2,4x,f6.1,4x,f5.1,4x,f5.1,4x,f6.0)
 1008 format (/,10x,'computation of rh from td',/)
c
c     Compute td from rh.
c
      write  (6,1009)
      write  (6,1010)
      do 6106 k = 1, nk
          tt      = t(k) + 273.16
          pp      = p(k)
          rrh     = rh(k)
          call TDFRMRH (pp,tt,rrh,td1)
          write (6,1011)k,pp,tt-273.16,rrh,td1-273.16
 6106 continue
 1009 format (/,10x,'computation of td from rh',/)
 1010 format (3x,'lev',3x,'pressure',1x,'tempera.',3x,'re hum',4x,
     &  'dew pt'./)
 1011 format (4x,i2,4x,f6.1,4x,f5.1,4x,f6.0,4x,f5.1)
c
c     Compute q from rh.
c
      write  (6,1012)
      write  (6,1013)
      do 6108 k = 1, nk
          tt      = t(k)+273.16
          pp      = p(k)
          rrh     = rh(k)
          call QFRMRH (pp,tt,rrh,qq)
```

```
          q(k)   = qq
          write (6,1014)k,pp,tt-273.16,rrh,qq
6108   continue
1012   format (/,10x,'computation of q from rh',/)
1013   format (3x,'lev',3x,'pressure',1x,'tempera.',3x,'re hum',4x,
   +   'sp hum',/)
1014   format (4x,i2,4x,f6.1,4x,f5.1,4x,f6.0,4x,f6.4)
c
c      Compute rh from q.
c
          write (6,1015)
          write (6,1016)
          do 6110 k = 1, nk
            tt        = t(k) + 273.16
            pp        = p(k)
            qq        = q(k)
            call rhfrmq (pp,tt,qq,rh1)
            write (6,1017)k,pp,tt-273.16,qq,rh1
6110   continue
1015   format (/,10x,'computation of rh from q',/)
1016   format (3x,'lev',3x,'pressure',1x,'tempera.',3x,'sp hum',4x,
   +   're hum',/)
1017   format (4x,i2,4x,f6.1,4x,f5.1,4x,f6.4,4x,f6.0)
          stop
          end
```

2. Determination of the Lifting Condensation Level (LCL)

The lifting condensation level is defined as the pressure level at which an air parcel reaches saturation by adiabatic cooling if it were raised vertically from the earth's surface. Its determination is relatively simple when the temperature, specific humidity and geopotential height are known at the surface. At the surface, the potential temperature is given by

$$\theta_s = T_s \left[\frac{1000}{P_s}\right]^{R/C_p} \tag{6.3}$$

where R and C_p represent the gas constant and the specific heat of air at constant pressure, respectively. Subscript 's' represents the surface level.

Table 6.1: Moisture variables conversions, P (mb), q (g/kg) and Rh (%).

P	T	q	T_d from q	q from T_d	R_h from T_d	T_d from R_h	q from R_h	R_h from q
1011.0	23.0	0.0170	22.5	0.0169	97.	22.5	0.0169	97.
1000.0	22.7	0.0164	21.8	0.0163	94.	21.7	0.0163	94.
955.4	22.0	0.0155	20.1	0.0154	89.	20.0	0.0154	89.
950.0	21.8	0.0155	20.0	0.0154	89.	20.0	0.0154	89.
900.0	18.4	0.0143	17.9	0.0142	97.	17.8	0.0142	97.
850.0	13.2	0.0116	13.2	0.0111	100.	13.1	0.0111	100.
816.2	15.6	0.0119	13.5	0.0118	87.	13.4	0.0118	87.
800.0	12.4	0.0112	12.4	0.0112	100.	12.3	0.0112	100.
782.4	11.3	0.0106	11.1	0.0105	98.	11.0	0.0105	98.
750.0	9.9	0.0101	9.7	0.0100	99.	9.7	0.0100	99.
722.4	8.8	0.0097	8.6	0.0096	98.	8.5	0.0096	98.
700.0	7.4	0.0091	7.2	0.0090	98.	7.1	0.0090	98.
687.1	6.0	0.0080	5.1	0.0080	94.	5.0	0.0080	94.
660.5	5.9	0.0068	2.2	0.0068	77.	2.1	0.0068	77.
655.1	5.7	0.0066	1.7	0.0066	75.	1.6	0.0066	75.
650.0	5.3	0.0064	1.1	0.0064	74.	1.1	0.0064	74.
641.2	4.6	0.0062	0.5	0.0062	75.	0.4	0.0062	75.
600.0	1.1	0.0055	-2.0	0.0055	79.	-2.1	0.0055	79.
594.3	0.6	0.0054	-2.4	0.0054	80.	-2.4	0.0054	80.
578.1	-0.6	0.0057	-2.0	0.0057	90.	-2.1	0.0057	90.

At the lifting condensation level P_l , T_l , q_l , and z_l are unknown. The determination of these variables is based on the potential temperature and dry static energy conservation principle. That is,

$$\theta_l = \theta_s \text{ and } C_p T_l + g\, Z_l = C_p T_s + g\, Z_s \tag{6.4}$$

The pressure at the lifting condensation level is obtained using an iterative scheme. A first guess pressure, P_l^1 , is chosen at 300 mb level. Poisson's equation is next solved for the temperature

$$T_l^1 = \theta_s \left[\frac{P_l^1}{1000} \right]^{R/C_p} \tag{6.5}$$

and a first guess geopotential height at the lifting condensation level is determined using the dry static energy conservation,

$$Z_l^1 = \{ Z_s + \frac{C_p}{g} (T_s - T_l^1) \} \tag{6.6}$$

The specific humidity may be then obtained as

$$q_l^1 = 0.622\, e_l^1 \ / \ (P_l^1 - 0.378\, e_l^1) \tag{6.7}$$

where

$$e_l^1 = 6.11 \exp \frac{a(T_l^1 - 273.16)}{T_l^1 - b} \tag{6.8}$$

and a, b are the constants defined in Section 1. If the iteration continues with an increment of 1 mb for the next guess on the pressure,

$$P_l^{v+1} = P_l^v + 1 \text{ mb} \tag{6.9}$$

one would find

$$q_l^{v+1} - q_s < 0 \tag{6.10}$$

This process is then continued until the inequality (6.10) first changes sign. At this point, the lifting condensation level is determined to within 1 mb of pressure and its height is calculated using (6.6).

Program (*LCL*), provided in this section, determines the pressure at the lifting condensation level. Sample outputs from this program are presented in Table 6.2.

```
        program L_C_L
c
c     This program detemines the lifting condensation level
c     given surface parameters. The lcl is obtained by
c     raising a parcel of air from the surface to a level
c     where it reachs saturation. The technique used is
c     based on the conservation of moist static energy.
c
c     Definition of surface parameters (pressure,temperature
c     and specific humidity)
c
        p0      = 1011.0
        t0      = 296.0
        q0      = 0.0170
c
c     Subroutine LCL determines the lcl.
c
        call LCL (p0,t0,q0,plcl,zlcl,tlcl,qlcl,n)
        write (6,1000)n
        write (6,1001)
        write (6,1002)tlcl,plcl,zlcl,qlcl
1000    format(10x,'number of itterations = ',i5,/)
1001    format(6x,'tlcl',4x,'plcl',5x,'zlcl',4x,'qlcl',/)
1002    format(3x,3f8.1,f8.4)
        stop
        end
```

3. Moist Adiabatic Profile

Another basic but important convection related topic in numerical weather prediction is the determination of the local moist adiabat. The moist adiabatic process is defined as the process under which a parcel of

Table 6.2: Determination of the lifting condensation level.

Number of iterations = 693

tlcl	plcl	zlcl	qlcl
295.4	993.0	60.8	0.0170

air undergoes its ascent after it reaches saturation. The moist adiabatic profile is commonly described by two invariants quantities, namely, the equivalent potential temperature and the moist static energy which are, respectively, expressed as

$$\theta_e = \theta \exp \left[\frac{Lq}{C_pT} \right] \tag{6.11}$$

where L is the latent heat of condensation and

$$E = C_pT + gz + Lq \tag{6.12}$$

Given the temperature, specific humidity and geopotential height at a reference pressure level P_r , the numerical problem is to determine the corresponding variables at any arbitrary pressure level P under the condition of moist static energy conservation. The entire procedure here assumes that the air parcel is saturated at the reference level; otherwise, the lifting condensation level must first be determined.

The algorithm proposed to determine these variables starts at a pressure level 1 mb lower than the reference pressure. At this level, an iterative scheme is carried out for temperature, and values of the specific humidity and geopotential height are calculated at each incremental iteration. The thickness between the two levels is estimated using the hydrostatic relation corrected with the virtual temperature.

$$\frac{\partial gz}{\partial p} = - \frac{R}{P} T_v \tag{6.13}$$

where the virtual temperature is given by

$$T_v = T (1. + 0.61q) \tag{6.14}$$

The moist static energy is then computed at this level and is compared to that of the reference pressure. The convergence is obtained when the moist static energy difference between the two levels is less than a fixed tolerance factor. Otherwise the temperature is incremented and the iteration is continued. The entire procedure is repeated for all levels. At the end of these calculations the values of temperature and specific humidity are determined along the moist adiabat whose profile can be constructed using (6.11). Two separate subroutines are provided to, respectively, estimate the profiles of temperature and specific humidity along the moist adiabat given N reference levels *MOIST* and to construct the moist adiabatic profile *THETAE*.

4. Convective Adjustment

Although unstable lapse rates may exist over some regions for a short period of time, they are not a persistent characteristic of the atmosphere. The existence of unstable lapse rates is undesirable during a numerical prediction. It leads to buoyancy driven convection and turbulence which can not be resolved by large scale numerical models. This sub-grid scale instability appears as noise to large scale models and can contaminate the forecast if it is not controlled. In numerical modeling, this problem may be prevented by the use of a strong vertical diffusion scheme or, more commonly, by convective adjustment. Convective adjustment consists essentially of replacing the unstable lapse rate by a neutral sounding, provided the total static energy of the column remains invariant. The treatment of this sub-grid scale instability is done separately for moist and dry atmospheric conditions.

4.1 Moist Convective Adjustment

Basically, moist convection is the procedure by which heating and moistening rates due to cumulus convection are determined in large scale numerical weather prediction models. It is invoked in regions of conditional instability associated with large scale upward motion and consists of replacing the prevailing unstable equivalent potential temperature profile by a neutral lapse rate. This energy redistribution procedure is based on the total moist static energy conservation in the adjusted layer. That is,

$$\int_{p_r}^{p_s} (C_pT + gz + Lq) \, dp = C^{te} \tag{6.15}$$

where p_r is a reference pressure level below which ($p \geq P_r$) the moist adjustment is performed. The adjustment procedure appears essentially as an extension of the moist adiabat construction principle. The only difficulty resides in the determination of the reference level. Once this level is obtained, the next step involves the calculation of vertically averaged values of the moist static energy E(p) from the surface ($p = p_s$) to various pressure levels, p, such as

$$\bar{E}(p) = \frac{\int_p^{p_s} (C_pT + gz + Lq) \, dp}{\int_p^{p_s} dp} \tag{6.16}$$

where p is less than the pressure at the level of minimum moist static energy (p_{min}). E(p) is tabulated at intervals of 10 mb starting from $p = p_{min} - 10$. At the same pressure level, E(p) would be very close to $\bar{E}(p)$, which determines the top pressure level p_r, for moist convective adjustment. In practice, moist convective adjustment is performed using the following steps.

i.   Interpolate the original sounding on to a fine resolution grid along the vertical coordinate. The determination of the reference pressure level becomes then much simpler. Program (*SPLUN*), which uses a cubic spline technique, is provided to perform such an interpolation from an original sounding on *NK* inequally spaced levels to *NI* levels. Other subroutines required to solve the linear system resulting from the cubic spline formulation are also provided. Figure 6.1 illustrates the use of *SPLUN*.

ii.   Using the interpolated profiles of temperature, geopotential and specific humidity, subroutine *ADJTOP* determines the top of the moist convective adjustment and returns the reference pressure level (p_r) with the corresponding reference height, temperature and specific humidity.

iii.   The construction of the adjusted sounding below this pressure level is then performed using subroutine *MOIST* in a reverse way; that is with a negative pressure increment. The number of levels to be adjusted must be carefully selected, NL = (p_s - p_r)/dp, where p_s is the surface pressure.

Program (*ADJUST*), provided below, uses interpolated profiles obtained from (*SPLUN*), and performs a moist convective adjustment. Results from this procedure are shown in Table 6.3.

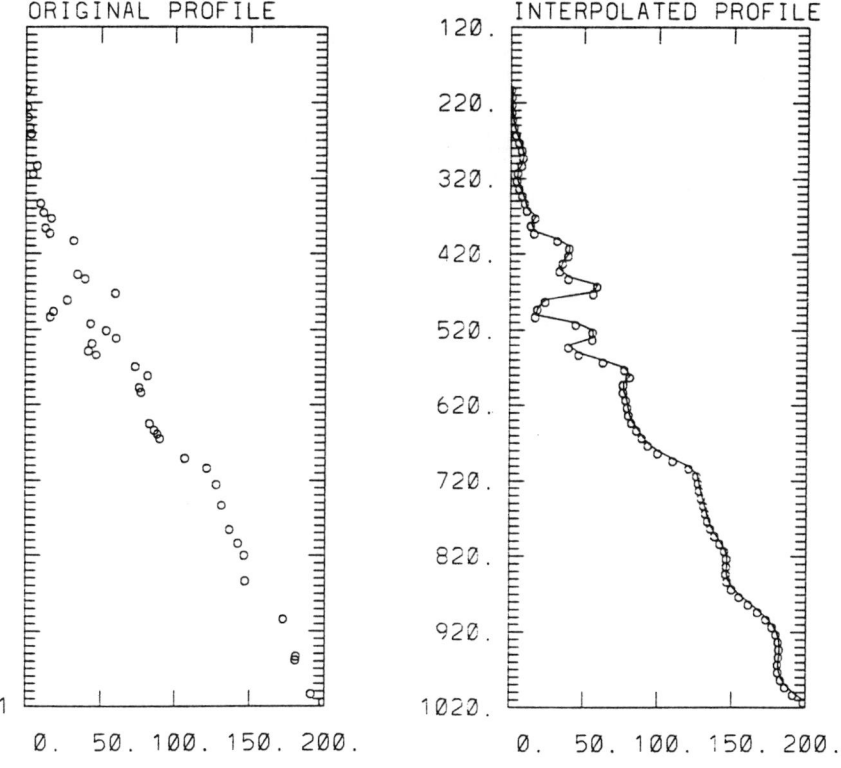

Figure 6.1: Interpolated specific humidity profile using the spline technique.

```
          program ADJUST
c
c     This program performs a moist convective adjustment.
c     The determination of the local moist adiabat is based
c     moist static energy conservation.
c
          parameter (n = 82)
          real t(n),q(n),p(n),z(n),wt(n),wp(n),wq(n)
          real t1(100),z1(100),q1(100),p1(n),wz(n)
c
c     Definition of constants
c
          data rgas,g/287.05, 9.81/
          rg        = rgas/g
c
c     Read input data from top to bottom.
c     file temp.dat contains: pressure and temperature
c     file spec.dat contains: pressure and specific humdity
c     file hght.dat contains: pressure and geopotential
c
          open (1,file='temp.dat',status='old')
          open (2,file='spec.dat',status='old')
          open (3,file='hght.dat',status='old')
c
          do 6300 k = 1, n
              read(1,101) p(k) ,t(k)
              read(2,102) p(k) ,q(k)
              read(3,102) p(k) ,z(k)
   6300   continue
    101   format(2f10.2)
    102   format(2f10.5)
c
c     Invert data to read from  bottom to top.
c
          do 6302 k = 1, n
              kk       = n-k+1
              wp(k)  = p(kk)
              wt(k)   = t(kk)
              wq(k)  = q(kk)
              wz(k)  = z(kk)
   6302   continue
          ps        = wp(1)
c
```

```
c       Determine the top of the adjustment
c
        call ADJTOP (wz,wt,wq,wp,p0,z0,t0,q0,n)
c
            if (p0.ne.0.) then
c
        write (6,1000)p0,t0,z0,q0
 1000   format(/,2x,'reference level                  ',f7.2,/,
     &          2x,'temperature at reference level   ',f7.2,/,
     &          2x,'height at reference level        ',f7.2,/,
     &          2x,'specific humidity at reference level ',f7.4  )
c
        nl       = int ((ps-p0)/10) + 1
        write (6,1001)nl
 1001   format(2x,'number of levels to be adjusted     ',i7,/)
c
c       Perform the adjustment of the layer below the
c       reference level. Note the negative pressure
c       increment.
c
        call MOIST (z0,t0,q0,p0,0.05,-10.,nl,z1,t1,q1,p1)
c
            endif
c
c       Display the original and the adjusted soundings
c       below the reference level.
c
        write (6,1002)
        write (6,1003)
        write (6,1004)
 1002   format(30x,'below adjustment top' ,/)
 1003   format(15x,'original sounding',22x,'adjusted sounding',/ )
 1004   format(4x,2('pressure',2x,'height',3x,'temp.',4x,'spec. hum.'
     &  ,2x),/)
 1005   format(4x,3(f7.2,2x),f7.4,5x,3(f7.2,2x),f7.4)
        do 6304 k = n-nl+1,n
            kk       = k-(n-nl)
            write (6,1005)p(k),z(k),t(k),q(k),p1(kk),z1(kk),t1(kk),q1(kk)
 6304   continue
        stop
        end
```

Table 6.3: Determination of the top of moist convective adjustment.

level	moist static energy (EM)	guess values for moist static energy (EADJ)	difference (EM − EADJ)
72	0.4131E+10	0.2497E+10	0.1634E+10
71	0.4095E+10	0.2411E+10	0.1684E+10
70	0.4061E+10	0.2367E+10	0.1694E+10
69	0.4027E+10	0.2363E+10	0.1663E+10
68	0.3992E+10	0.2372E+10	0.1620E+10
67	0.3956E+10	0.2366E+10	0.1590E+10
66	0.3921E+10	0.2357E+10	0.1563E+10
65	0.3884E+10	0.2412E+10	0.1472E+10
64	0.3847E+10	0.2322E+10	0.1525E+10
63	0.3810E+10	0.2313E+10	0.1497E+10
62	0.3772E+10	0.2525E+10	0.1248E+10
61	0.3731E+10	0.2617E+10	0.1114E+10
60	0.3687E+10	0.2564E+10	0.1123E+10
59	0.3644E+10	0.2457E+10	0.1187E+10
58	0.3602E+10	0.2380E+10	0.1221E+10
57	0.3560E+10	0.2422E+10	0.1138E+10
56	0.3517E+10	0.2657E+10	0.8595E+09
55	0.3468E+10	0.2562E+10	0.9064E+09
54	0.3421E+10	0.2064E+10	0.1357E+10
53	0.3382E+10	0.1977E+10	0.1404E+10
52	0.3344E+10	0.1896E+10	0.1448E+10
51	0.3307E+10	0.2229E+10	0.1078E+10
50	0.3262E+10	0.2318E+10	0.9444E+09
49	0.3215E+10	0.2267E+10	0.9476E+09
48	0.3168E+10	0.2013E+10	0.1154E+10
47	0.3125E+10	0.2057E+10	0.1068E+10
46	0.3080E+10	0.2194E+10	0.8858E+09
45	0.3031E+10	0.2306E+10	0.7257E+09
44	0.2979E+10	0.2292E+10	0.6870E+09
43	0.2926E+10	0.2189E+10	0.7369E+09
42	0.2874E+10	0.2136E+10	0.7373E+09
41	0.2821E+10	0.2102E+10	0.7195E+09
40	0.2769E+10	0.2057E+10	0.7118E+09
39	0.2716E+10	0.2011E+10	0.7052E+09
38	0.2663E+10	0.1974E+10	0.6896E+09
37	0.2610E+10	0.1951E+10	0.6588E+09

Table 6.3: Continued

level	moist static energy (EM)	guess values for moist static energy (EADJ)	difference (EM − EADJ)
36	0.2556E+10	0.1927E+10	0.6287E+09
35	0.2501E+10	0.1901E+10	0.5992E+09
34	0.2445E+10	0.1895E+10	0.5493E+09
33	0.2387E+10	0.1918E+10	0.4697E+09
32	0.2327E+10	0.1946E+10	0.3814E+09
31	0.2265E+10	0.1922E+10	0.3425E+09
30	0.2200E+10	0.1865E+10	0.3356E+09
29	0.2136E+10	0.1803E+10	0.3335E+09
28	0.2072E+10	0.1745E+10	0.3272E+09
27	0.2007E+10	0.1689E+10	0.3179E+09
26	0.1942E+10	0.1636E+10	0.3066E+09
25	0.1877E+10	0.1580E+10	0.2965E+09
24	0.1811E+10	0.1520E+10	0.2911E+09
23	0.1745E+10	0.1456E+10	0.2893E+09
22	0.1679E+10	0.1417E+10	0.2612E+09
21	0.1611E+10	0.1428E+10	0.1836E+09
20	0.1540E+10	0.1444E+10	0.9567E+08
19	0.1464E+10	0.1418E+10	0.4543E+08
18	0.1385E+10	0.1311E+10	0.7437E+08
17	0.1308E+10	0.1093E+10	0.2145E+09
16	0.1240E+10	0.1241E+10	-0.1435E+07

reference level   850.0000
temperature   288.8000
height   1596.000
specific humidity   0.1465000
number of levels to be adjusted = 17

4.2 Dry Convective Adjustment

The notion of dry convective adjustment simply describes the effect of energy dissipation in dry atmospheric conditions. Dry convection occurs over regions where the lapse rate is greater than the dry adiabatic value, and transports energy upward in order to neutralize the instability of the layer. The potential energy is transformed into kinetic energy which is then dissipated into heat. This adjustment in the stability of the layer is done under the constraint of total energy conservation. There exists a pressure level, p_r, above the level of minimum dry static energy, which determines the top of the convective adjustment. At this level, if the observed dry static energy is E_r, then

$$E_r(p_s - p_r) = \int_{p_r}^{p_s} E\, dp \qquad (6.17)$$

where E is the dry static energy of the sounding and p_s represents the surface pressure assumed to be 1000 mb, in this case. The procedure consists of integrating the dry static energy E from a level $p_r = p_s - n\Delta p$ to the pressure level p_s, starting at n=1. This integration continues until the left hand side of (6.17) equals or exceeds the right hand side. At this point, the adjusted sounding is then recovered using the conservation of dry static energy, Poisson equation and the hydrostatic law. The adjustment cools the lower layers and warms the upper layers of the adjusted sounding. If successive layers are unstable, or if a layer becomes unstable following the adjustment of the layer immediately below it, they are treated simultaneously to produce a neutral stability in the combined layer. In practice, the adjustment is done over successive layers, in which the potential temperature is used to determine the stability. As an unstable layer is found, the average dry static energy is computed, and at each level of the layer the adjusted constant potential temperature is estimated using the mean dry static energy for that layer. This can be formalized as

$$E = C_pT + gz \qquad (6.18)$$

and

$$\theta = T\left[\frac{P_o}{P}\right]^\kappa \quad \text{with} \quad \kappa = \frac{R}{C_p} \qquad (6.19)$$

The derivative of (6.18) with respect to p, yields

$$\frac{\partial E}{\partial p} = g\frac{\partial z}{\partial p} + C_p\frac{\partial T}{\partial p} \qquad (6.20)$$

Using the hydrostatic relation, (6.20) may be expressed as

$$\frac{1}{C_p}\frac{\partial E}{\partial p} = \frac{\partial T}{\partial p} - \frac{RT}{C_pP} \qquad (6.21)$$

On the other hand, it is easy to see that

$$\frac{\partial \theta}{\partial p} = \left[\frac{P_0}{P}\right]^\kappa \left[\frac{\partial T}{\partial p} - \frac{RT}{C_p P}\right]$$

(6.22)

Combining (6.21) and (6.22), the following equation giving the potential temperature as a function of the dry static energy is obtained,

$$\frac{\partial \theta}{\partial p} = \frac{1}{Cp} \left[\frac{P_0}{P}\right]^\kappa \frac{\partial E}{\partial p}$$

(6.23)

Subroutine *DCONADJ* which uses this formulation to perform a dry convective adjustment is provided and amply documented.

5. A Simple Cloud Model

In the moist convective adjustment concept, several assumptions were implicitly made in that the complex physical processes related to precipitation were not invoked and all the condensation produced by the process was supposed to precipitate as rain. The nature of precipitation within a cloud is extremely complex and is yet not completely monitored in large scale numerical models.

In this section, a construction of a simple cloud model is described. Specifically, a buoyancy driven cloud model presented by Nickerson (1965) is discussed and a numerical algorithm simulating the heat source effect on the thermals during the integration of the basic hydrodynamical equations is provided. The model's governing equations are not fully discussed in this book and further details concerning the physical process involved should be sought from the original paper. The framework of this model is based on the two-dimensional shallow Boussinesq system. The motion is assumed nondivergent, and the basic equations considered in the vertical (x,z)-plane reduce to

$$u = \frac{\partial \psi}{\partial z}$$

(6.24)

$$w = -\frac{\partial \psi}{\partial x}$$

(6.25)

The vorticity and heat transfer equations are expressed as

$$\frac{\partial \xi}{\partial t} = J(\psi,\xi) - g \frac{\partial \varphi}{\partial x} + \nu \nabla^2 \xi \tag{6.26}$$

and

$$\frac{\partial \varphi}{\partial t} = J(\psi,\varphi) - \frac{Q}{\theta_o} + \nu \nabla^2 \varphi \tag{6.27}$$

where the initial state potential temperature anomaly is given by

$$\varphi = \frac{\theta - \theta_o}{\theta_o} \tag{6.28}$$

and θ_o represents a constant potential temperature defining an initially undisturbed neutral environment. The horizontal component of the vorticity is related to the streamfunction as

$$\xi = \nabla^2 \psi \tag{6.29}$$

A buoyant element defined by

$$\varphi(x,z,\theta) = 0.5 \cos \left[\frac{\pi x}{320} \right] \cos^2 \left[\frac{\pi(z-100)}{400} \right] \tag{6.30}$$

is initially introduced into a neutral environment in the region comprised between 0 and 160 m in the horizontal direction and between 100 and 300 m in the vertical.

The total plane extends 760 meters along the x-direction and 600 meters in the vertical. The symmetry of the buoyant element helps reduce the computational domain to only half the plane in the x-direction.

Furthermore, a sustained heating is defined over the region $0 \le x \le 160$ and $80 \le z \le 120$. This diabatic heating source maintains a buoyancy at the bottom of the growing cloud element and is defined by

$$Q = Q_o \cos \left[\frac{\pi x}{320} \right] \cos^2 \left[\frac{\pi(z-100)}{400} \right] \tag{6.31}$$

This model is regarded as a slab symmetric model where at x = 0 the symmetry condition imposes

$$\frac{\partial w}{\partial x} = \frac{\partial \psi}{\partial x} = 0 \tag{6.32}$$

Furthermore, at z = 0 and z = 600 m, the following conditions are applied: $w = 0$, $\dfrac{\partial u}{\partial x} = 0$ and $\dfrac{\partial \varphi}{\partial z} = 0$. The lateral derivatives $\dfrac{\partial w}{\partial x}$ and $\dfrac{\partial \varphi}{\partial x}$ are set to zero at $x = 380$ m, and the zonal wind component is set to zero at $x = 0$ and $x = 380$ m.

A complete code describing the construction of this simple shallow convection cloud model is provided and is amply documented. All the subroutines used in this program (*NICKER*) have already been introduced and discussed in previous chapters. Outputs illustrating the evolution of the cloud at different times during the integration are presented in Figures 6.2, 6.3, and 6.4.

program *NICKER*

```
c
c      This program simulates the evolution of a simple
c      cloud model as described by Nickerson (1965).
c      The heat source is inserted at the bottom of the
c      atmosphere and the fluid is initially stratified
c      and at rest.
c      The thermal begins to ascent,and a vortex circulation
c      develops . The temperature within the thermal
c      increases due to heating and the thermal rises above the
c      heat source. The maximum temperature reaches a peak
c      then decreases when dissipation exceed buoyancy.
c      The potential temperature field resembles a thin
c      stemmed mushroom.
c
c      Define variables.
c
c      n        : vertical dimension
c      m        : horizontal dimension
c      dt       : time step here dt = 3 seconds
c      loopt    : output interval equivalent to 2 minutes
c      loop     : number of time steps. here max 200 = 10 minutes
c      gnu      : eddy kinematic coefficient of heat and momentum
c      dx       : grid spacing. 10 meters in horizontal and vertical
c      tneut    : potential temperature of neutral environment
c      alpha    : relaxation factor
c      eta      : vorticity (per sec)
c      t        : temperature excess (deg celcius)
c      psi      : streamfunction
c      difu     : eddy dissipation (m**2/sec)
```

```
c
        parameter (l = 39, m = 61,n = 39)
        common d(61),tmap(39,61),wrk(39,61)
        common psi(39,61),eta(39,61,2),t(39,61,2),q(39,61),
     &          arak(39,61),difu(39,61),u(39,61),w(39,61)
c
        open(20,file='nicker.out',status='unknown')
c
  16    format(4x,'the number of scans required to obtained the',
     &   1x,'following stream function was ',i3,' in this time step')
  17    format(4x,'excess potential temperature at
     &   time= ',f5.1,' seconds')
  18    format(4x,'stream function',10x,'elasped time= ',f5.1,
     &   'seconds')
  19    format(4x,'vorticity',15x,'time=',f5.1)
  20    format(4x,'vertical velocity',15x,'time',f5.1)
  21    format(4x,'horizontal velocity',15x,'time=',f5.1)
  25    format(4x,'the number of scans required for the first'
     &   ,'relaxation in this time step was ',i3)
 1001   format(/)
        write(6,1001)
c
c    Define constants
c
        n1        = n-1
        m1        = m-1
        n2        = n-2
        m2        = m-2
        dt        = 3.
        loopt     = 40
        gnu       = 0.5
        dx        = 10.
        e         = dx
        tneut     = 300.
        g         = 9.81
        pi        = 4.0*atan(1.0)
        nold      = 1
        new       = 2
        alpha     = 1.89

        do 6600 i = 1, m
 6600   d(i)    = dx
c
```

```
c     Initialize eta,phi,psi,and q
c
      call BASIC (eta(1,1,nold),t(1,1,nold),psi,
     &                   u,w,q,n,m,n1,m1,tneut)
      time      = 0.

      do 6602 i = 1, 1
      do 6602 j = 1, m
 6602 wrk(i,j)= t(i,j,nold)
c
c     Write output of initial state.
c
      write (20,119) time
      write (20,120) wrk
  119 format(f12.2)
  120 format(6e13.6)

      do 6604 i = 1, 16
      do 6604 j = 10, 30
      t(i,j,nold) = t(i,j,nold)/tneut
 6604 continue
      loop      = 0
c
c     Begin new time loop
c
  999 continue
      if (loop.eq.200) go to 35
      loop      = loop+1
      if (loop.eq.1) go to 1
      nsave     = nold
      nold      = new
      new       = nsave
    1 continue
c
c     Define the heating as a function of time
c
      do 6608 j = 9,13
      do 6608 i = 1,17
         qq     = 4.e-03*cos(pi*(i-1)/32.)*
     &            (cos(pi*((j-1)-10.)/4.))**2
      q(i,j)    = 2.0*qq/pi
      if (loop.gt.100.and.loop.le.200) q(i,j) = -2.0*qq/pi
 6608 continue
```

```
c
c     Compute initial estimate of eta from vorticity equation.
c     Routine JAC computes the horizontal advection of vorticity.
c     Routine LAPLAC computes the Laplacian of the streamfunction.
c
          call JAC        (arak,psi,eta(1,1,nold),d,e,n,m,n1,m1,n2.m2)
          call LAPLAC (eta(1,1,nold),difu,dx,n,m,n1,m1)
          do 6610 i = 2, n1
          do 6610 j = 2, m1
             eta(i,j,new) = eta(i,j,nold)+dt*arak(i,j)+dt*gnu*difu(i,j)
     &                      -dt*g*(t(i+1,j,nold)-t(i-1,j,nold))/(2.*dx)
 6610   continue
          do 6612 i = 1, n
             eta(i,1,new) = 0.
             eta(i,m,new) = 0.
 6612   continue
          do 6614 j = 1, m
             eta(1,j,new) = 0.
             eta(n,j,new) = 0.
 6614   continue
c
c     Calculate initial estimate of phi from the
c     heat transfer equation.
c
          call JAC        (arak,psi,t(1,1,nold),d,e,n,m,n1,m1,n2,m2)
          call LAPLAC (t(1,1,nold),difu,dx,n,m,n1,m1)
          do 6616 i = 1, n
          do 6616 j = 1, m
             t(i,j,new) = t(i,j,nold)+dt*arak(i,j)+dt*q(i,j)
     &                      /tneut+dt*gnu*difu(i,j)
 6616 continue
c
c     Relax eta to get psi.Routine RELAX1 solves
c     the poisson equation.
c
          call RELAX1 (psi,eta(1,1,new),n,m,nscan,alpha)
c
c     The horizontal and the vertical velocity
c     are defined from the streamfunction.
c
          do 6618 i = 1, n
          do 6618 j = 1, m1
             u(i,j)  = (psi(i,j+1)-psi(i,j))/dx
```

```
 6618    continue
         do 6620 j = 1, m
         do 6620 i = 1, n1
            w(i,j)  = -(psi(i+1,j)-psi(i,j))/dx
 6620    continue
c
c     Calculate final (corrector) estimates of eta and phi for this
c     time step from the predicted vorticity and temperature fields.
c
         call JAC       (arak,psi,eta(1,1,new),d,e,n,m,n1,m1,n2,m2)
         call LAPLAC (eta(1,1,new),difu,dx,n,m,n1,m1)
         do 6622 i = 2, n1
         do 6622 j = 2, m1
         eta(i,j,new) = eta(i,j,nold)+dt*arak(i,j)+dt*gnu*difu(i,j)
      &                  -dt*g*(t(i+1,j,new)-t(i-1,j,new))/(2.*dx)
 6622    continue
c
c     Relax final estimate of eta to get final psi
c     and consequently u and w fields.
c
         call RELAX1 (psi,eta(1,1,new),n,m,nscan,alpha)
         do 6624 i = 1, n
         do 6626 j = 1, m1
            u(i,j)   = (psi(i,j+1)-psi(i,j))/dx
 6626    continue
 6624    continue
         do 6628 j = 1, m
         do 6628 i = 1, n1
            w(i,j)  = -(psi(i+1,j)-psi(i,j))/dx
 6628    continue
c
c     Compute final estimate of phi for this time step
c
         call JAC       (arak,psi,t(1,1,new),d,e,n,m,n1,m1,n2,m2)
         call LAPLAC (t(1,1,new),difu,dx,n,m,n1,m1)
         do 6630 i = 1, n
         do 6630 j = 1, m
            t(i,j,new) = t(i,j,nold)+dt*arak(i,j)
      &          +dt*q(i,j)/tneut+dt*gnu*difu(i,j)
 6630    continue
c
c     Check if output is required.
c     Loopt is the number of time steps equivalent to 2 minutes
```

```
c       for output interval. The outputs can be displayed on the
c       screen by calling cloud with the fourth argument being 1.
c
        time       = time+dt
        if (mod(loop,loopt).ne.0) go to 999
        write (6,16)nscan
        time       = 3.0*loop
        write (6,17)time
        call CLOUD (t(1,1,new),n,m,0)
        do 6632 i = 1, n
        do 6632 j = 1, m
        tmap(i,j)  = t(i,j,new)*tneut
 6632   continue
        shditv     = 0.1

        do 6634 j = 1, m
        do 6634 i = 1, 1
 6634   wrk(i,j)   = tmap(i.j)
c
c       Write output of temperature anomaly
c       for each required time.
c
        write (20,119)time
        write (20,120)wrk
c
        write (6,18)time
        call CLOUD (psi,n,m,0)
        shditv     = time/60.
c
c       Write output of streamfunctions
c       for each required time.
c
        write (20,119)time
        write (20,120)psi

        write (6,19)time
        call CLOUD (eta(1,1,new),n,m,0)
        shditv     = time/30000.
        write (6,20)time
        do 6636 j = 1, m
            w(n,j) = w(n1,j)
 6636   continue
        call CLOUD (w,n,m,0)
```

```
        shditv     = time/1200.
        write (6,21)time
        do 6638 i = 1, n
            u(i,m) = u(i,m1)
6638    continue
        call CLOUD (u,n,m,0)
        if (loop.eq.300) go to 35
        go to 999
35      continue
        stop
        end
```

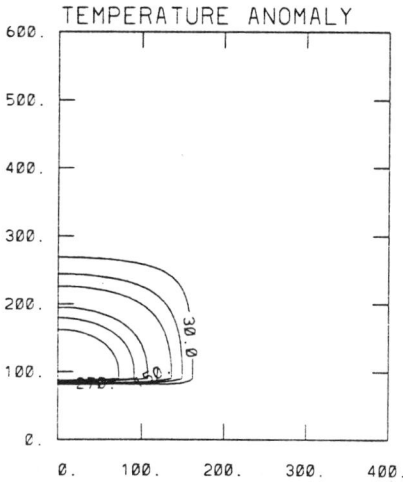

Figure 6.2: Vertical cross-section of the initial temperature anomalies (degrees). Scaling factor 10^3.

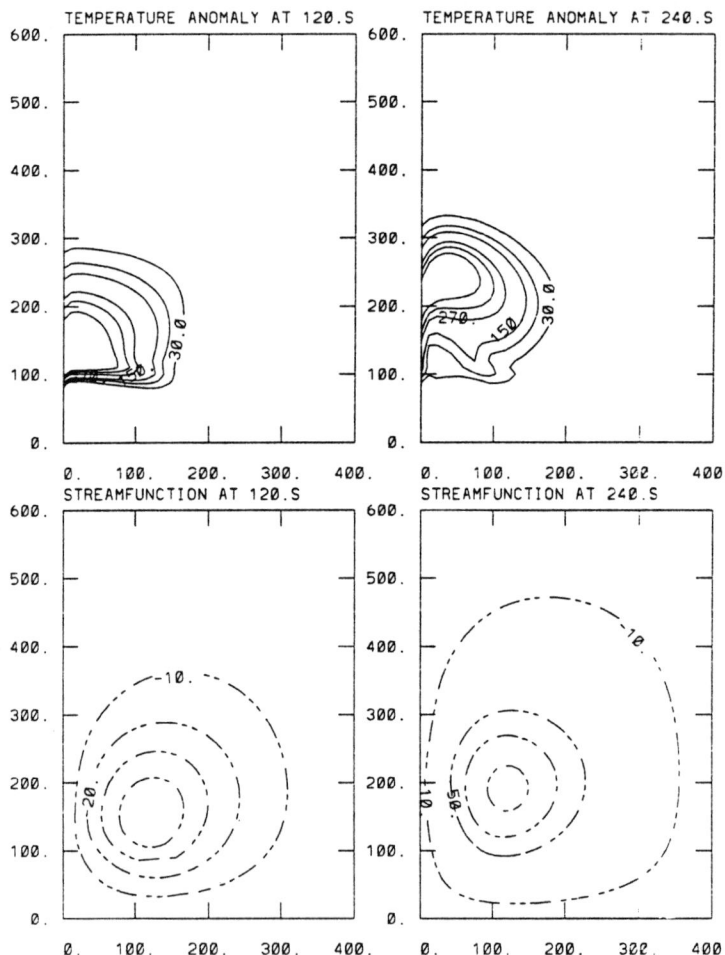

Figure 6.3: Vertical cross-section of the temperature anomalies and
streamfunctions evolution after 2 and 4 min of
integration. Scaling factor 10^3 for temperature and 10^8
for streamfunctions. (Dashed contours are negative.)

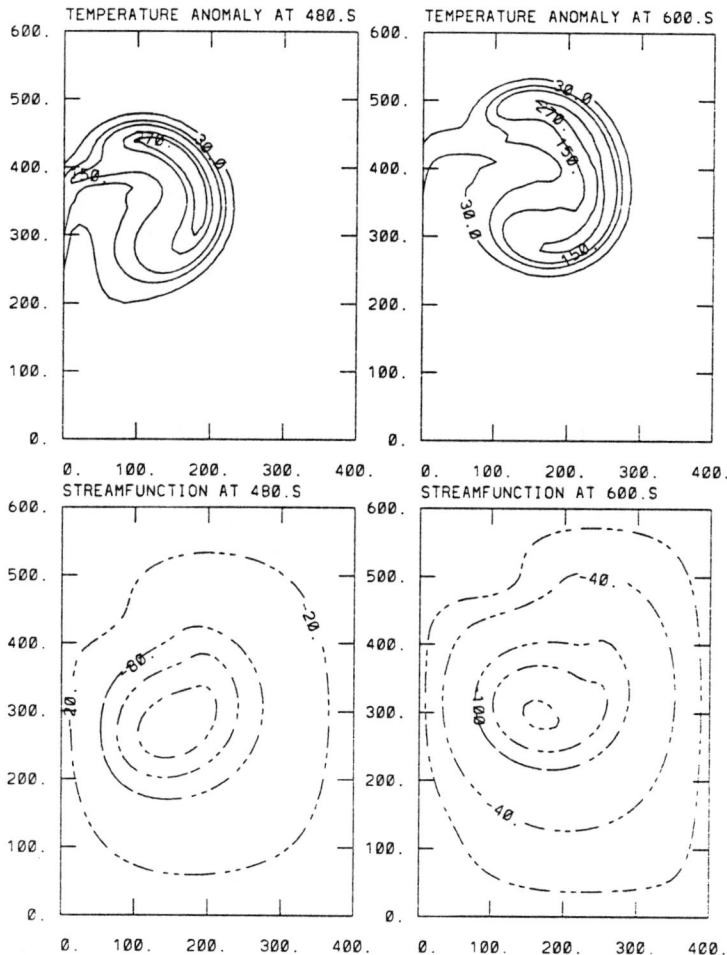

Figure 6.4:   Temperature anomalies and streamfunctions evolution
after 8 and 10 min of integration. Scaling factor 10^3
for temperature and 10^8 for streamfunctions.

7

Cumulus Convection and Large Scale Condensation

Cumulus convection represents one of the central physical processes in numerical weather prediction. Nevertheless, despite intensive research during the last two decades, the description of convective clouds in numerical weather prediction is still not completely adequate. This appears mostly related to the complex structure of convective clouds and their mutual interactions with the environment in which they evolve. Although the spatio-temporal scale of cumulus clouds is about two orders of magnitude less than the synoptic scale, observations have shown that their organization in clusters may have a significant influence on the large scale atmospheric motion through heat, momentum, and moisture exchanges. Furthermore, because of the complex mechanisms taking place inside convective clouds, such as ice formation, melting, evaporation, sublimation and precipitation, cumulus clouds are not explicitly described in large scale numerical models. Only their mean statistical effects on the large scale circulation is parameterized.

In this chapter, some of the most commonly used schemes for the determination of the heating, moistening and rainfall rates resulting from cumulus convection are discussed. Most importantly, a detailed code carrying out these calculations is provided and illustrated. Calculation of the stable heating during large scale condensation is also presented.

1. Cumulus Convection

1.1 Kuo Scheme

The convection scheme introduced by Kuo (1965) basically describes the effect of condensation and its consequent large scale apparent heating and moistening on a tropical cyclone. This scheme assumes cumulus convection to occur in deep conditionally unstable layers experiencing a net low level moisture convergence. The model cloud is defined by a local moist adiabat whose base is at the condensation level of surface air and whose top is limited at the level where this moist adiabat intercepts the environmental sounding. Since its introduction in 1965, this scheme has undergone several modifications.

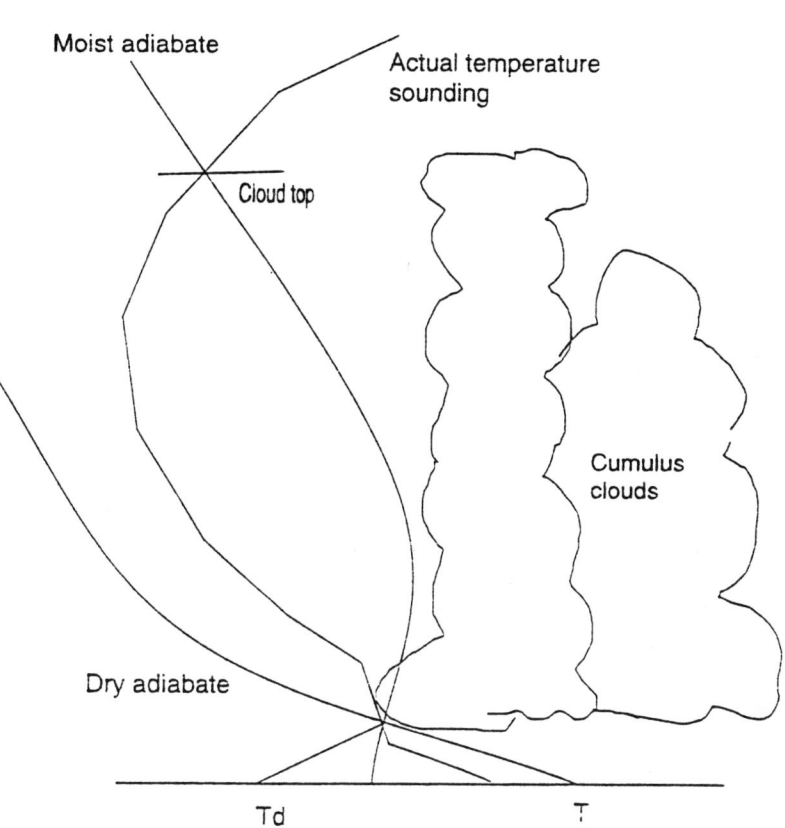

Figure 7.1: Schematic cloud base and top levels.

1.1.1 Original Kuo Scheme

In its original formulation, Kuo's scheme described the moisture equation as

$$\frac{\partial q}{\partial t} = a \frac{q_s - q}{\Delta \tau} \tag{7.1}$$

where $\Delta \tau$ is the cloud time scale and the parameter a denotes the fractional area that would be covered by newly generated convective cells. Here q_s represents the saturation specific humidity and q is the specific humidity along the environmental sounding. The fractional area is defined as

$$a = \frac{-\frac{1}{g} \int_{P_T}^{P_B} (\nabla \cdot q\vec{V} + \frac{\partial}{\partial p} q\omega) \ dp}{\frac{1}{g} \int_{P_T}^{P_B} \left[\frac{C_p(T_s - T)}{L\Delta\tau} + \frac{q_s - q}{\Delta\tau} \right] dp} \quad (7.2)$$

where p_T and p_B are the pressures at the top and bottom of the cloud, respectively. In this relation, the denominator may be interpreted as the total amount of water vapor supply required to cover the entire grid box area by a model cloud. The numerator, on the other hand, represents the available moisture supply. This available moisture is used to create a model cloud (a local moist adiabat) at (T_s, q_s) from the environmental sounding at (T, q). Part of this moisture is assumed to condense and precipitate, contributing then to an increase in the air temperature from T to T_c. The remaining moisture goes necessarily to moisten the atmosphere, bringing its specific humidity from q to q_s.

The convective rainfall rate is defined as

$$P_t = \frac{1}{g} \int_{P_T}^{P_B} \frac{a \ C_p(T_s - T)}{L\Delta\tau} \ dp \quad (7.3)$$

The consistency of this scheme with respect to moisture conservation can be shown by considering the moisture conservation law,

$$\frac{dq}{dt} = E - P \quad (7.4)$$

where E and P denote the total evaporation and precipitation rates, respectively. Therefore,

$$\frac{\partial q}{\partial t} = - \nabla . q \overrightarrow{V} - \frac{\partial}{\partial p} q\omega + E - P \tag{7.5}$$

Integrating from p_T to p_B and assuming that evaporation occurs only below the cloud base, (7.5) becomes

$$\frac{1}{g} \int_{p_T}^{p_B} \frac{\partial q}{\partial t} \, dp = - \frac{1}{g} \int_{p_T}^{p_B} (\nabla . q \overrightarrow{V} + \frac{\partial}{\partial p} q\omega) \, dp - \frac{1}{g} \int_{p_T}^{p_B} P \, dp \tag{7.6}$$

Introducing

$$Q = \frac{1}{g} \int_{p_T}^{p_B} \left[\frac{C_p (T_s - T)}{L\Delta\tau} + \frac{q_s - q}{\Delta\tau} \right] dp \tag{7.7}$$

(7.6) may be rewritten as

$$\frac{1}{g} \int_{p_T}^{p_B} \frac{\partial q}{\partial t} \, dp = I - p_t \tag{7.8}$$

where

$$I = aQ \tag{7.9}$$

This yields

$$\frac{1}{g} \int_{p_T}^{p_B} \frac{\partial q}{\partial t} \, dp = \frac{a}{g} \int_{p_T}^{p_B} \frac{q_s - q}{\Delta\tau} \, dp \tag{7.10}$$

If one takes the moisture conservation law and integrates it between the cloud base (lifting condensation level) and the cloud top (where the buoyancy vanishes), one can show that it reduces to the above equation provided one makes use of the definition of the net moisture convergence in vertical columns. Furthermore, the moisture convergence may be divided into two parts such as

$$I = I_\theta + I_q \qquad\qquad (7.11)$$

where I_θ and I_q represent the heating and moisting parts, respectively.

Several studies have shown the inadequacy of this scheme in large scale applications. It was found that the available moisture supply, I, was not proportionally distributed between heating and moistening rates. Most of the supply was found to be used in moistening, causing then an early saturation prior to the establishment of a neutral moist adiabatic sounding on the large scale. Furthermore, the heating rates and the rainfall totals were usually underestimated in cases where the horizontal resolution was about a few hundred kilometers.

1.1.2 Modified Kuo Scheme

Kuo (1974) introduced some modifications to the original cumulus parameterization scheme. It was suggested that the warming due to adiabatic compression in downdraft regions of the cloud are implicitly included in the scheme. Moreover and most importantly, the partitioning of the total moisture supply introduced was such that a fraction b is used to moisten the air while the remainder portion $(1 - b)$ is either used to produce rain or to contribute to the warming of the air. It was also suggested that over low level convergence tropical regions the fraction $(1 - b)$ was dominant. Nevertheless, no formal framework was suggested for the parameter b, and its determination was empirical. Anthes (1977) suggested an inverse relationship between the fractional moistening rate b and the mean relative humidity, implying more moistening in drier environments. However, no theoretical solution was proposed for the moistening parameter.

1.1.3 FSU Scheme

Basically the Florida State University (FSU) cumulus parameterization algorithm is an extension of the Kuo scheme in which the closure problem has been solved via statistical regressions. Generally, Kuo's cumulus parameterizations define the moisture supply from the vertical integral of the moisture flux large scale convergence given by (in advective form)

$$C_M = -\bar{V}.\bar{\nabla}\bar{q} - \bar{\omega}\frac{\partial\bar{q}}{\partial p} \qquad (7.12)$$

This supply serves as a basis for the definition of cloud elements in the Kuo type of schemes. Krishnamurti et al. (1983) and Anthes (1977) noted that only the second term of (7.12) accounts for the definition of clouds and that the horizontal moisture advection contributes directly to the moistening of the large scale. Therefore, the mositure supply is defined as

$$I_L = -\frac{1}{g}\int_{P_T}^{P_B} \omega\frac{\partial q}{\partial p}\, dp \qquad (7.13)$$

where p_T and p_B denote the cloud's top and base, respectively, as defined in Section 1.1.

Semiprognostic studies using the GARP* Atlantic Tropical Experiment (GATE) observations (Krishnamurti et al., 1980, 1983) revealed that the definition of the moisture supply as suggested by (7.13) is a close measure of the rainfall rate and, thus, is not sufficient to account for the observed moistening of the vertical columns. An introduction of a mesoscale convergence parameter was then suggested to account for this moisture deficit. This lead to a definition of the moisture supply as

$$I = I_L (1 + \eta) \qquad (7.14)$$

where the term, $I_L \eta$, represents the mesoscale moisture supply. The total moisture supply is then partioned into precipitation and moistening using the following relations:

$$R = I(1 - b) = I_L(1 + \eta)(1 - b) \qquad (7.15)$$

and

$$M = Ib = I_L(1 + \eta)b \qquad (7.16)$$

Following Krishnamurti et al. (1983), the total moisture necessary to produce a grid-scale moist adiabatic sounding is defined as

---

*GARP: **G**lobal **A**tmospheric **R**esearch **P**rogram

$$Q = \frac{1}{g} \int_{P_T}^{P_B} \frac{q_s - q}{\Delta \tau} \, dp + \frac{1}{g} \int_{P_T}^{P_B} \left[\frac{c_p T(\theta_s - \theta)}{L\theta \Delta \tau} + \omega \left[\frac{c_p T}{L\theta} \right] \frac{\partial \theta}{\partial p} \right] dp \qquad (7.17)$$

which may be written as

$$Q = Q_q + Q_\theta \qquad (7.18)$$

Here $\Delta \tau$ denotes the cloud time scale and is approximately equal to 30 minutes. Likewise, the total moisture supply may be partitioned into moistening and heating parts such as

$$I_q = Ib = I_L \, b(1 + \eta) \qquad (7.19)$$

and

$$I_\theta = I(1 - b) = I_L(1 - b)(1 + \eta) \qquad (7.20)$$

These effects are introduced in large scale numerical models by expressing the thermodynamic and the moisture equations as

$$\frac{\partial \theta}{\partial t} + \vec{V}.\nabla \theta + \omega \frac{\partial \theta}{\partial p} = a_\theta \left[\frac{\theta_s - \theta}{\Delta t} + \omega \frac{\partial \theta}{\partial p} \right] \qquad (7.21)$$

and

$$\frac{\partial q}{\partial t} + \vec{V}.\nabla q = a_q \left[\frac{q_s - q}{\Delta t} \right] \qquad (7.22)$$

where

$$a_\theta = I_\theta / Q_\theta = \frac{I_L(1+\eta)(1-b)}{Q_\theta} \qquad (7.23)$$

and

$$a_q = I_q / Q_q = \frac{I_L b(1+\eta)}{Q_q} \qquad (7.24)$$

The terms on the right hand side of (7.21) and (7.22) represent the cumulus effect.

Since Q_θ and Q_q are known, the parameterization is closed if b and η are determined. Using GATE data, Krishnamurti et al. (1983) proposed a closure for b and η based on a statistical approach. Normalized heating (R/I_L) and moistening (M/I_L) were expressed as functions of large scale variables using multiple regressions. Screening of the large scale variables showed that the vertically integrated vertical velocity and the 700 mb relative vorticity were the two most important predictors. The heating and moistening are then expressed as

$$R / I_L = a_1 \xi + a_2 \overline{\omega} + a_3 \tag{7.25}$$

and

$$M / I_L = b_1 \xi + b_2 \overline{\omega} + b_3 \tag{7.26}$$

where a_1, a_2, a_3, b_1, b_2, b_3 are the regression coefficients and may be found in Krishnamurti et al. (1983).

In numerical weather prediction, ξ and ω determine M / I_L and R / I_L. The moistening parameters b and η are next determined using (7.15) and (7.16). Finally a_θ and a_q are obtained from

$$a_\theta = \frac{I_L (1-b)(1+\eta)}{Q_\theta} = R / Q_\theta , \tag{7.27}$$

and

$$a_q = \frac{I_L (1+\eta)b}{Q_q} = M / Q_q \tag{7.28}$$

The large scale apparent heat source and moisture sink are then expressed, respectively, as

$$Q_1 = a_\theta \left[c_p \frac{T}{\theta} \frac{\theta_s - \theta}{\Delta \tau} + \omega c_p \frac{T}{\theta} \frac{\partial \theta}{\partial p} \right] + c_p \frac{T}{\theta} (H_R + H_s) \tag{7.29}$$

and

$$Q_2 = - L \left[a_q \frac{q_s - q}{\Delta \tau} + \omega \frac{\partial q}{\partial p} \right] \tag{7.30}$$

where H_R and H_s represent the radiative and sensible heating rates, respectively. The total convective precipitation is finally obtained as

$$P_T = \frac{1}{g} \int_{P_T}^{P_B} a_\theta \, c_p \, \frac{T}{\theta} \, \frac{\theta_s - \theta}{\Delta \tau} \, dp \qquad (7.31)$$

The terms Q_1 and Q_2 represent the heating and moistening rates due to subgrid-scale phenomena (diabatic effects), which are apparent to the large scale atmospheric motion.

In this chapter, a Fortran code computing the heating, moistening and rainfall rates for the generalized Kuo scheme is provided. Illustrations of the apparent heat source and moisture sink at 500 mb as computed by program (*KUO*) are presented in Figures (7.2) and (7.3).

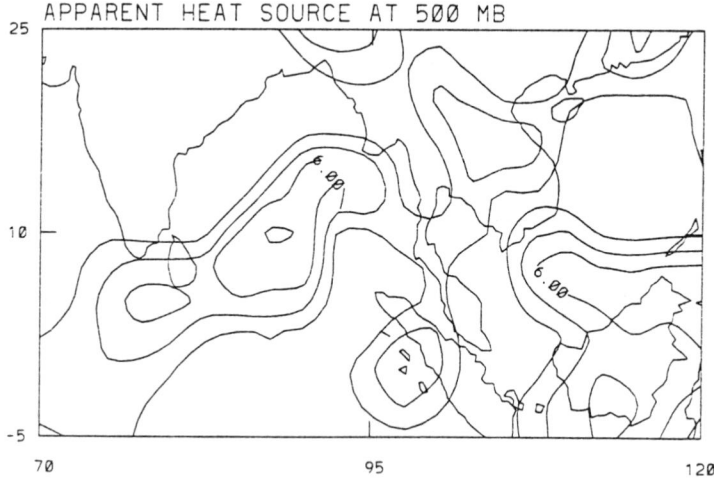

Figure 7.2: Apparent heating due to cumulus clouds (deg/day).

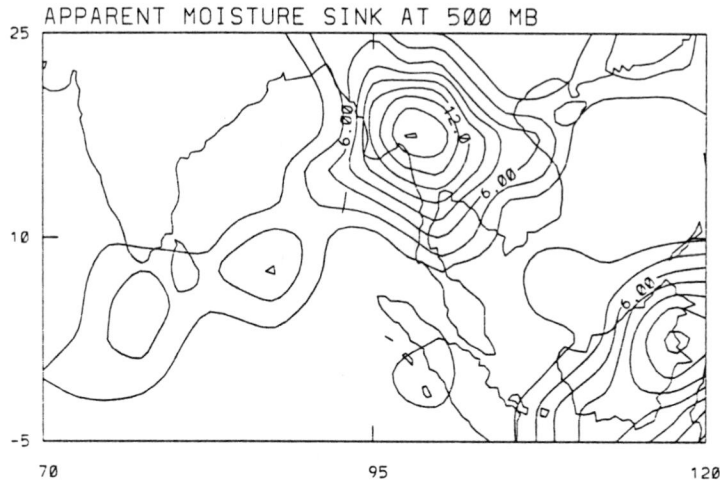

Figure 7.3: Apparent moistening due to cumulus cloud (deg/day).

The corresponding rainfall rates integrated over the entire column is shown in Figure (7.4).

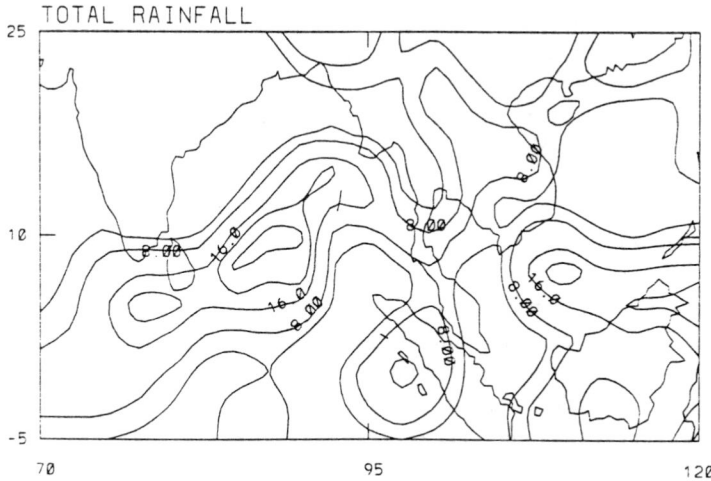

Figure 7.4: Rainfall rates due to cumulus clouds (mm/day).

The vertical structure of the heat source and moisture sink for one selected point within the domain is also shown in Figure (7.5).

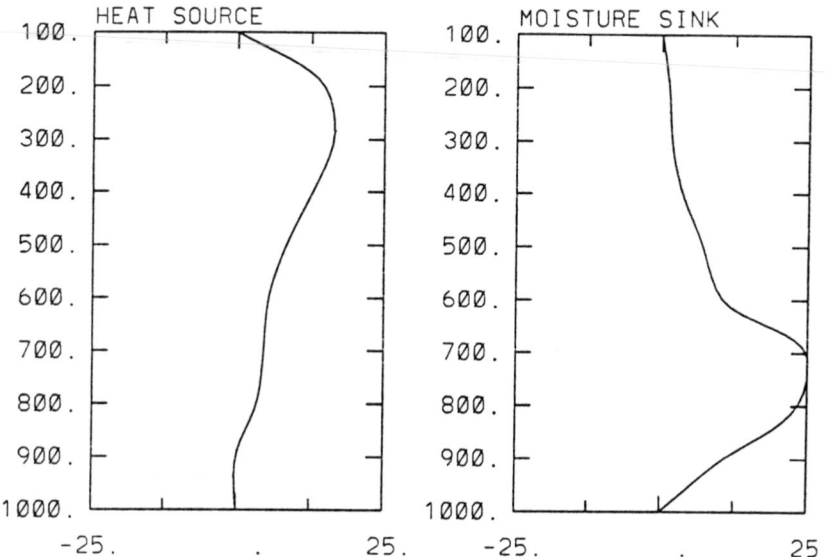

Figure 7.5: **Vertical profiles of the heating and moistening due to cumulus clouds (deg/day).**

The grid point illustrated in Figure (7.5) is arbitrarily chosen and is located at 62.5 east and 5 north. The rainfall rate is given by the vertical integral of the heating rate expressed in water column units. Program (*KUO*) can be easily manipulated to obtain the heating and moistening rates of the classical Kuo scheme.

program *KUO*

```
c
c     This program computes cloud related heat source (q1),
c     moisture sink(q2), and rainfall rate for a vertical
c     column over a given grid point.
c     This column model requires input of temperature,
c     dew point temperature and vertical velocity at n
c     levels in the vertical. It also requires the 700 mb
c     relative vorticity at that point. The moistening
c     parameter,b and the meso-scale convergence, eta are
```

```
c       computed before the main subroutine cvheat is called.
c
c
c       Definitions of variables
c
c       dp          : pressure increment           (mb  )
c       tp          : temperature                  (K   )
c       td          : dew point temperature        (K   )
c       w           : vertical p-velocity          (mb/s )
c       wbar        : average vertical velocity
c       vort        : vorticity at 700 mb          (sec-1)
c       pt          : pressure at temperature levels (mb  )
c
        parameter (n = 10)
        real pt(n),t(n),td(n),w(n) ,q1(n), q2(n)
c
c       Input data.
c
        data pt/1000.,900.,800.,700.,600.,500.,400.,300.,200.,100./
        data t/301.46,296.45,290.51,285.24,279.28,270.57,258.77,
     +       243.39,222.21,189.53/
        data td/298.79,294.64,286.67,275.86,268.12,260.73,251.75,
     +       241.81,220.63,187.96/
        data w/-0.16250e-02,-0.17760e-02,-0.18789e-02,-0.20905e-02,
     +       -0.21773e-02,-0.18447e-02,-0.13510e-02,-0.10983e-02,
     +       -0.91707e-03,-0.30293e-03/
c
        open(2,file ='prof.dat',status='unknown')
c
        vort        = 0.81442e-05
        dp          = 100.
c
c       Define coefficients of multiple regression for
c       heating and moistening parameters
c
        a           = 0.158e05
        b           = 0.304e03
        c           = 0.476
        p           = 0.107e05
        q           = 0.107e03
        r           = 0.870
c
c       Compute the average vertical velocity
```

```
            wbar      = 0.
            do 7100 k = 1, n
                wbar  = wbar + w(k)/float(n)
    7100 continue
c
c     Compute b and eta
c
            rnumer    = a*vort+b*wbar+c
            denom     = (a+p)*vort+(b+q)*wbar+(c+r)
            yb        = rnumer/denom
            eta       = denom-1.
c
c     Reset parameter b to [0,1] in case it is out of range.
c     this can happen when running the code over large domain.
c     the coeffiecients given here were obtained using GATE
c     domain only.
c
            if (yb.gt.1.0) yb = 1.0
            if (yb.lt.0.0)  yb = 0.0
c
c     Call the main routine to compute the heating ,moistening
c     and rainfall rates.
c
            call CVHEAT (pt,t,td,w,yb,eta,n,dp,q1,q2,rrr)
c
c     Display outpts for the column
c
            do 7102 k = 1, n
                print *,q1(k),q2(k),rrr
                write (2,1000)q1(k),q2(k),rrr
    7102    continue
    1000    format (2x,3e12.5)
            stop
            end
```

2. Arakawa-Shubert Cumulus Parameterization Scheme

In 1974, Arakawa and Shubert introduced a scheme which describes the theory of interaction between a cumulus cloud ensemble and the large scale environment. The theory formulates the control of the large scale motion on cumulus ensemble. Basically, large scale heat and moisture budget equations are developed for regions above and within the subcloud mixed layer. This large scale forcing is based on a quasi-equilibrium of the cloud work function, which represents an integral measure of the buoyancy force of the cumulus clouds defined for each cloud type (Arakawa and Schubert, 1974). Similarly to other schemes, cumulus clouds modify the large scale temperature and moisture through detrainment, condensation products and subsidence. The detrainment induces large scale cooling and moistening, while the subsidence causes warming and drying. The unique feature in this scheme is the partitioning of cumulus ensemble into cloud types defined by a fractional entrainment rate λ. The normalized cloud vertical mass flux and the cloud physical properties are controlled by the large scale environment. This is referred to as the dynamic control (Arakawa and Schubert, 1974). The quasi-equilibrium hypothesis assumes that the rate at which large scale processes generate conditional instability is balanced by the rate at which clouds suppress conditional instability through feedback. This method is described in detail in the original paper by Arakawa and Schubert (1974). However, the most relevant point in terms of numerical calculations is the solution of the integral equation defining the cloud base mass flux. Once this is determined, the total mass flux as well as the heating and moistening rates may be obtained without computational difficulties.

In the Arakawa-Schubert cumulus parameterization scheme, the cloud base mass flux is expressed in the form of an integral equation of the type,

$$I(p) = \int_0^{\lambda_D} M_B(\lambda) \, A(\lambda) \, d\lambda \tag{7.32}$$

where $M_B(\lambda)$ is the cloud base mass flux, $A(\lambda)$ is the cloud work function, λ denotes a particular cloud type and λ_D is a known function of pressure. Since clouds can in principle have any possible heights, $p \, (\lambda_D)$ will theoretically range between $p=0$ and $p=1000$ mb. It should be noted that in (7.32), $I(p)$ and $A(\lambda)$ are known functions and the only unknown is the cloud base mass flux $M_B(\lambda)$.

The atmosphere is then divided into n layers of depth Δp each, such that $n\Delta p = 1000$ mb and the shallowest cloud's top is at $(1000 - \Delta p)$. For this cloud, the integral equation is written as

$$I(1000 - \Delta p) = \int_0^{\lambda_D(1000-\Delta p)} M_B(\lambda) \, A(\lambda) \, d\lambda \qquad (7.33)$$

The right hand side of (7.33) can be thought of as a sum over $(n-1)$ layers, and the integral may then be represented by a finite summation containing $(n-1)$ unknowns $M_B(\lambda)$. This can be written as a linear algebraic equation of the type,

$$I(1000-\Delta p) = A_1 M_B(\lambda_1) + A_2 M_B(\lambda_2) + \ldots + A_{n-1} M_B(\lambda_{n-1}) \qquad (7.34)$$

The integral equation for clouds reaching the level $(1000 - 2\Delta p)$ is then recurrently written as

$$I(1000-2\Delta p) = B_1 M_B(\lambda_1) + B_2 M_B(\lambda_2) + \ldots + B_{n-2} M_B(\lambda_{n-2}) \qquad (7.35)$$

Sequentially, $(n-1)$ equations can be formed for the cloud base mass fluxes. This system is described by a triangular matrix and can be efficiently solved by backward substitution for the entire integral equation. It should be noted, however, that there are many difficulties in maintaining the positive definiteness of the cloud base mass flux. There are various programming methods and additional constraints invoked to preserve this condition. Once $M_B(\lambda)$ is determined the total mass flux is obtained by integration over all cloud types.

It was shown that in the presence of moist convection, the large scale atmospheric thermodynamic structure strongly couples changes in temperature to changes in specific humidity. Large scale apparent heat source, moisture sink and precipitation rates were also found to be adequately described by the quasi-equilibrium hypothesis.

3. Large Scale Condensation

As discussed in Chapter 3, it is of great importance to include the large scale condensation heating among the physical processes governing the atmospheric circulation. Under the condition that large scale can not be super-saturated, large scale condensation heating also called stable heating may be estimated from the time rate of change of the saturation specific humidity. In numerical weather prediction models, stable heating is considered only if the following three conditions are simultaneously satisfied

    i.   Saturated or near-saturated atmosphere, $q/q_s \approx 1$

    ii.   Existence of upward vertical motion, $\omega < 0$

    iii.   Absolutely stable atmosphere, $-\dfrac{\partial \theta_e}{\partial p} > 0$

Under these conditions the stable heating rate is defined as

$$H_s = -L \frac{dq_s}{dt} \tag{7.36}$$

where q_s is the saturation specific humidity and L is the latent heat of condensation. In regions of ascending motion, the stable heating is approximated by

$$H_s = -L\omega \frac{\partial q_s}{\partial p} \tag{7.37}$$

The term $\dfrac{\partial q_s}{\partial p}$ actually represents the slope of the saturation humidity along the local reference moist adiabatic which passes through the point defined by P, T, q_s and where the heating is calculated. At this reference point, the moist static energy is expressed as

$$E_m = C_p T + gz + Lq_s \tag{7.38}$$

The conservation of the moist static energy along the local moist adiabat leads to

$$C_p \frac{\partial T}{\partial p} - \frac{RT}{p}(1 + 0.61q_s) + L\frac{\partial q_s}{\partial p} = 0 \tag{7.39}$$

Thus,

$$\frac{\partial T}{\partial p} = \frac{RT}{C_p P}(1 + 0.61q_s) - \frac{L}{C_p}\frac{\partial q_s}{\partial p} \tag{7.40}$$

From Teten's formula, the saturation specific humidity may be approximated by

$$q_s \approx \frac{0.622e_s}{P} \tag{7.41}$$

where

$$e_s = 6.11 \exp \frac{a(T-273.16)}{(T-b)} \tag{7.42}$$

and

$$P \gg e_s \tag{7.43}$$

Therefore,

$$\frac{\partial q_s}{\partial p} = -\frac{0.622 e_s}{p^2} + \frac{0.622}{p} \frac{\partial e_s}{\partial p} \tag{7.44}$$

Let

$$E^* = \exp \frac{a(T-273.16)}{(T-b)} \text{ , then}$$

$$\frac{\partial q_s}{\partial p} = \frac{-0.622 \times 6.11}{p} \left[\frac{E^*}{p} - \frac{\partial E^*}{\partial p} \right] \tag{7.45}$$

where

$$\frac{\partial E^*}{\partial p} = \frac{\partial E^*}{\partial T} \times \frac{\partial T}{\partial p} = E^* \left[\frac{a}{T-b} - \frac{a}{(T-b)^2} \frac{(T-273.16)}{} \right] \frac{\partial T}{\partial p} \tag{7.46}$$

Substituting for $\frac{\partial T}{\partial p}$ from (7.40), a final equation is obtained for $\frac{\partial q_s}{\partial p}$ as

$$\frac{\partial q_s}{\partial p} = -\frac{0.622 \times 6.11}{P} \exp \frac{a(T-273.16)}{T-b} \left[\frac{1}{p} - \left\{ \frac{a}{T-b} \right. \right.$$

$$\left. \left. - \left[\frac{a(T-273.16)}{(T-b)^2} \right] \right\} \left\{ \frac{RT}{C_pP} \left[1 + 0.61 q_s \right] - \frac{L}{C_p} \frac{\partial q_s}{\partial p} \right\} \right] \tag{7.47}$$

The total stable rainfall rate is then obtained as

$$R_s = \frac{1}{g} \int_0^{P_s} \frac{H_s}{L} \, dp \tag{7.48}$$

It is generally more convenient to express the heating rate in units of °C day^{-1} and the rainfall rate in mm day^{-1}. In this case, it is noted that

$$\frac{H_s}{C_p} = -\frac{L\omega}{C_p} \frac{\partial q_s}{\partial p} \tag{7.49}$$

has the units of $^\circ C\ s^{-1}$. Hence,

$$\frac{86400\ \ H_s}{C_p} = \frac{86400}{C_p}\ \frac{L\omega}{C_p}\ \frac{\partial q_s}{\partial p} \tag{7.50}$$

is expressed in $^\circ C\ day^{-1}$. On the other hand, if ω is expressed in mbs^{-1}, g in ms^{-2} and q_s in gm/gm, then

$$R_s = -\frac{86400}{g}\int_0^{P_s}\omega\ \frac{\partial q_s}{\partial p}\ dp \tag{7.51}$$

would provide the total rainfall rate in mm day^{-1}.

Subroutine **STBHEAT** is provided to carry out these calculations and returns the stable heating in units of $^\circ C\ day^{-1}$. It should be kept in mind that the stable heating is zero if any of the above mentioned three conditions is not met.

8

Planetary Boundary Layer

This chapter addresses the computation of the different energy fluxes exchanged at the earth-atmosphere interface, namely the fluxes of momentum, sensible and latent heat. These fluxes result from the partitioning of the net solar radiation absorbed at the ground. At this point, the net radiative flux at the surface is assumed to be known and reference be made to Chapter 9 for its computation. In the following sections, the so-called bulk aerodynamic as well as the surface similarity theory formulations are discussed in the context of surface fluxes estimation. Land and oceanic regions are treated separately.

1. Bulk Aerodynamic Calculation over Ocean and Land

In the bulk aerodynamic formulation, the surface fluxes of momentum, sensible and latent heat, are commonly expressed as

$$F_h = C_p \rho \, C_h \, |V_a| \, (T_s - T_a) \tag{8.1}$$

$$F_q = \rho \, L \, C_q \, |V_a| \, (q_s - q_a) \tag{8.2}$$

$$F_m = \rho \, C_d \, |V_a| \, |V_a| \tag{8.3}$$

Here F_m, F_h, and F_q denote the fluxes of momentum, sensible and latent heat, respectively. V_a is the wind speed and T_a is the air temperature at the anemometer level. T_s and q_s denote the surface temperature and the saturation specific humidity, respectively. Over water, T_s and q_s are replaced by sea surface temperature T_w and saturation specific humidity q_w at that temperature, respectively. C_p is the specific heat of air at constant pressure, ρ is the air density and L represents the latent heat of condensation. C_h, C_q, and C_d are the nondimensional turbulent exchange coefficients and have the currently accepted values

$$C_h = 1.4 \text{ x } 10^{-3} \qquad \text{turbulent exchange coefficient of heat} \qquad (8.4)$$

$$C_q = 1.6 \text{ x } 10^{-3} \qquad \text{turbulent exchange coefficient of moisture} \qquad (8.5)$$

$$C_d = 1.1 \text{ x } 10^{-3} \qquad \text{turbulent exchange coefficient of momentum} \qquad (8.6)$$

These values were obtained from a field experiment during GATE (1974).

The choice of units is quite important in the estimation of these fluxes. For most meteorological purposes, it is desirable to express F_h and F_q in units of watts/m^2, while F_m is usually expressed in units of Dynes/cm^2. Using these units, and $\rho = 1.23 \text{ x } 10^3$ g/cm^3, the bulk formulae reduce to

$$F_h = 1.73 \left| V_a \right| (T_w - T_a)$$

$$F_q = 4.9 \text{ x } 10^3 \left| V_a \right| (q_w - q_a) \qquad (8.7)$$

$$F_m = 1.35 \text{ x } 10^{-2} V_a^2$$

where the wind speed is measured in m sec^{-1}, the temperature in degrees centigrade and the specific humidity in g kg^{-1}.

In situations of strong surface winds, for example, in tropical storms and hurricanes, or when dealing with ocean case where waves exert a large drag on the air, it is desirable to allow for a variation of the drag coefficient as a function of the wind speed. This functional relationship is described in Section 2, related to the surface roughness parameter calculation. Subroutine *BLKFLX*, provided in this section, estimates the surface fluxes using the bulk aerodynamic method with a wind dependent drag coefficient.

2. The Roughness Parameter

Although recent investigations on land surface processes have shown a strong interdependence between surface characteristics such as albedo, surface roughness, soil moisture and the vegetative cover, this section restricts the discussion of the roughness parameter to two simple cases, namely, the ocean and bare land.

To perform global calculations of the roughness parameter, a land-sea matrix as well as the surface height field are required. Assume that the land-sea matrix has values of

$$NTYPE = 0 \qquad \text{over ocean}$$
$$\text{and} \qquad NTYPE = -1 \qquad \text{over land}$$

and that a field of orographic height is prescribed for the entire domain of grid-points.

2.1 Ocean Case

Over oceans, the Charnock (1955) formula is an accepted method for defining the roughness length. It is expressed as

$$Z_0 = M \, u_*^2/g \tag{8.8}$$

where M is a constant having a value of 0.04 and u_* is the friction velocity and is a function of Z_0. Therefore an iterative procedure is necessary for the determination of the roughness parameter. Starting from

$$u_*^2 = (-u'w') = \tau_0 / \rho_0 \tag{8.9}$$

where τ_0 is the surface stress and ρ_0 is the air density, a first guess for u_* can be obtained from the bulk aerodynamic representation provided the variation of the drag coefficient as a function of the wind speed is allowed. In (8.9) the prime represents deviation from the mean and the eddy covariance term $u'w'$ represents the momentum flux of the zonal wind component. Accepted values of the drag coefficients may be represented as

$$\begin{aligned}
C_d &= 1.1 \times 10^{-3} & &\text{for } V < 5.8 \text{ m sec}^{-1} \\
C_d &= (0.74 + 0.046V) & &\text{for } 5.8 < V < 16.8 \text{m sec}^{-1} \quad (8.10) \\
C_d &= (0.94 + 0.034V) & &\text{for } V > 16.8 \text{m sec}^{-1}
\end{aligned}$$

The first guess field of Z_0 is obtained from the observed wind speed V, using the following steps.

$$V \rightarrow C_d \rightarrow (C_d \ V^2) \rightarrow u_* \rightarrow Z_0$$

Using this first guess for Z_0, the similarity approach (discussed below) is used to determine the surface fluxes including u_*, which in turn defines the final value of the roughness parameter.

2.2 Land Case

Over land, the suggested method is defined following Delsol et al. (1981) which allows for a variation of the roughness parameter as a function of the ground elevation. The method is based on the mesoscale variance of mountain heights. In its simplest form it is expressed as

$$Z_0 = 15 + (473.6 + 0.03684 \ h)^2 \times 10^{-6} \qquad (8.11)$$

where the grid scale mountain height, h, and the roughness parameter, Z_0, are expressed in cm. For numerical weather prediction, it is desirable to restrict the upper limit of Z_0 to 4000 cm. If (8.11) is used for the estimation of roughness length in a numerical model, its computation can be carried out separately and treated as an input for the surface flux estimation algorithm. The iterative method applied over oceans may be found in subroutine *SFLX*. An example of land-sea matrix as well as a topographic height field for a limited domain are shown as examples in Figure (8.1). These two fields are useful for the calculation of the roughness length.

3. Surface Fluxes from Similarity Theory

In many planetary boundary layer problems the physics is too complex and is insufficiently known to formulate principles for the governing equations. Planetary boundary layer observations constitute, therefore, the only basis for any formulation used in the estimation of the surface fluxes. The similarity theory is based on such observations and relates the non dimensionalized vertical gradient of the large scale wind, potential temperature and specific humidity to the nondimensional height scale (z/L) using Buckingham Pi dimensional analysis (Businger et al.,

LAND-SEA MASK

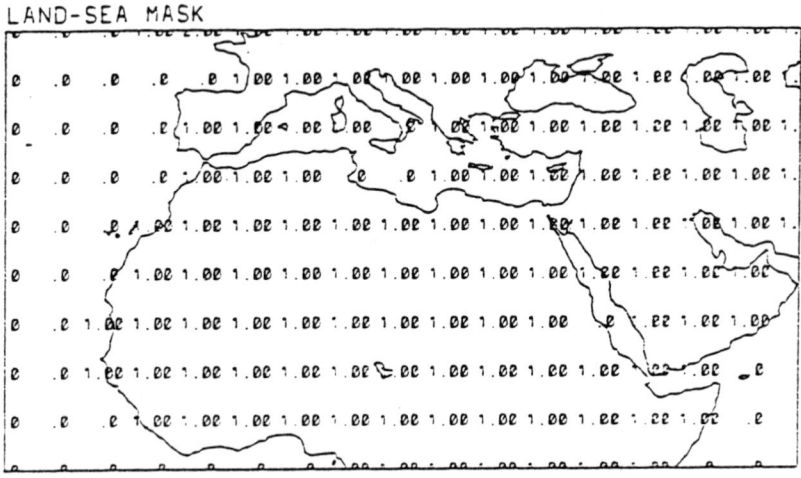

TOPOGRAPHY FIELD

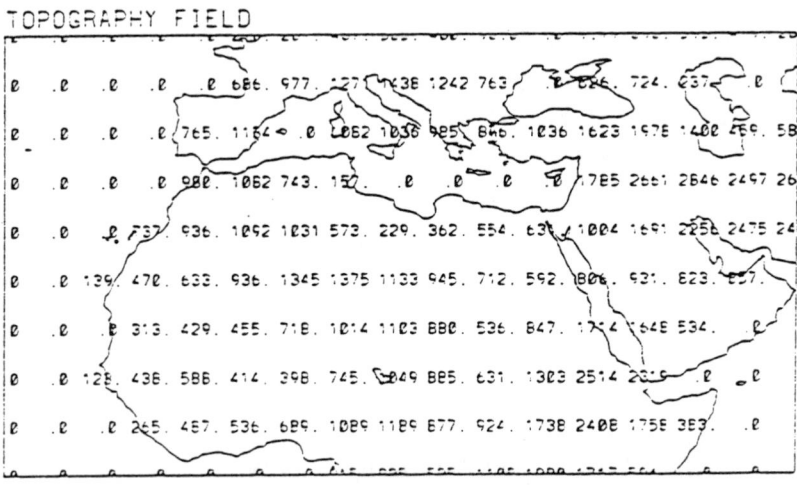

Figure 8.1: Example of land-sea matrix (dimensionless) and topography
            fields (meters).

1971). Here z is the height above the ground level and L is the
Monin-Obukhov length defined as

$$L = u_*^2 \,/\, k\beta\theta_*$$ (8.12)

where β is the buoyancy parameter and is given by $\beta = g/\bar{\theta}$. $\bar{u}_*$ and $\bar{\theta}$ represent the friction velocity and a reference temperature, respectively. The application of the similarity principle to the nondimensional wind shear, thermal gradient and specific humidity gradient gives

$$\frac{kz}{u_*} \frac{\partial \bar{u}}{\partial z} = \phi_m (z / L) \tag{8.13}$$

$$\frac{kz}{\theta_*} \frac{\partial \bar{\theta}}{\partial z} = \phi_h (z / L) \tag{8.14}$$

$$\frac{kz}{q_*} \frac{\partial \bar{q}}{\partial z} = \phi_q (z / L) \tag{8.15}$$

where $k = 0.35$ (Businger et al., 1971) is the Von Kármán constant and θ_* and q_* are the characteristic potential temperature and specific humidity, respectively. The terms ϕ_m, ϕ_h, ϕ_q are the nondimensional functions for vertical gradients of momentum, heat, and moisture, respectively. These functional forms were obtained by empirical best fit curve from observations of the nondimensional wind shear, potential temperature, and specific humidity gradients to the nondimensional length scale. The fits of the planetary boundary layer observations are generally treated in terms of stability. Since, a priori the Monin-Obukhov length is unknown, the bulk Richardson number, $RiB = g\Delta\theta z / \bar{\theta}u^2$, is used to determine the stability of the surface layer. RiB is positive for the stable case and negative for the unstable case. Following Bussinger et al. (1971), the treatment is done separately for stable and unstable cases.

This theory has been extensively documented in the meteorological literature, and it is well supported by observations. The description of the surface fluxes presented in this section follows the analysis of Louis (1979).

The integration of the flux profile relationships between two levels z_1 and z_2 assumed both to be contained in the constant flux layer leads to

$$u(z_2) = u(z_1) + \frac{u_*}{k} \left[\ln(z_2/z_1) - \phi_m (z_2/L) + \phi_m (z_1/L) \right] \tag{8.16}$$

and

$$\theta(z_2) = \theta(z_1) + R \frac{\theta_*}{k} \left[\ln(z_2/z_1) - \phi_h \, (z_2/L) + \phi_h \, (z_1/L) \right] \qquad (8.17)$$

ϕ_m and ϕ_h denote universal functions of z/L, where z_1 is taken as the roughness length, z_0, while z_2 is taken as the lowest model level. The constant R, ratio of the drag coefficients for momentum and heat in the neutral case was estimated to be 0.74 (Bussinger et al., 1971). Using this notation and assuming $u(z_0) = 0$, (8.16) and (8.17) may be rewritten as

$$u(z) = \frac{u_*}{k} \left[\ln(z/z_0) - \phi_m \, (z/L) + \phi_m \, (z_0/L) \right] \qquad (8.18)$$

and

$$\theta(z) = \theta_0 + R \frac{\theta_*}{k} \left[\ln(z/z_0) - \phi_h \, (z/L) + \phi_h \, (z_0/L) \right] \qquad (8.19)$$

where θ_0 is the potential temperature at z_0 considered the same as the ground temperature. The velocity, temperature and height scales are defined in terms of the vertical eddy fluxes of momentum, sensible heat and the Monin-Obukhov length, respectively, as

$$u_* = \left| u'w' \right|^{\frac{1}{2}} \qquad (8.20)$$

$$\theta_* = - \overline{w'\theta'} \, / \, u_* \qquad (8.21)$$

and

$$L = \overline{\theta u_*^2} \, / \, kg\theta_* \qquad (8.22)$$

It should be noted that a relation similar to (8.19) may be obtained for the specific humidity variable. Following Bussinger et al. (1971), the flux profile relationships are represented as

$$\phi_m(\mu) = \ln \left[\left[\frac{1+x}{2} \right]^2 \left[\frac{1+x^2}{2} \right] \right] - 2\arctan x + \frac{\pi}{2} \qquad \text{stable conditions} \qquad (8.23)$$

$$\phi_m(\mu) = -\beta_m \, \mu \qquad \qquad \text{unstable conditions} \qquad (8.24)$$

and

$$\phi_h(\mu) = \ln \left[\frac{1+y}{2} \right] \qquad \qquad \text{stable conditions} \qquad (8.25)$$

$$\phi_h(\mu) = -\beta_h \ \mu/R \qquad\qquad \text{unstable conditions} \qquad (8.26)$$

where $x = (1 - \gamma_m \ \mu)^{\frac{1}{2}}$, $y = (1 - \gamma_h \ \mu)^{\frac{1}{2}}$, $\beta_m = \beta_h = 4.7$, $\gamma_m = 15.0$, $\gamma_h = 9.0$ (Bussinger et al., 1971). The determination of the eddy fluxes of momentum and heat requires the elimination of the Monin-Obukhov length from (8.18) and (8.19) using (8.12). In stable conditions the procedure appears straightforward, but in unstable conditions a transcendental equation arises which can only be solved numerically. For computational efficiency, Louis (1979) proceeded as follows. Substituting (8.17) and (8.18) into (8.12), an implicit relationship between the bulk Richardson number, R_iB, and the Monin-Obukhov length was developed as

$$L = \frac{\overline{\theta}u^2}{g\Delta\theta} \frac{\ln(z/z_0 \ - \ \phi_m(z/L) \ + \ \phi_h(z_0/L)}{[\ln(z/z_0 \ - \ \phi_m(z/L) \ + \ \phi_m(z_0/L)]^2} \qquad (8.27)$$

Formal relationships were then obtained for the momentum and heat fluxes as

$$u_*^2 = a^2 \ u \ F_m \ (z/z_0 \ , \ R_iB) \qquad (8.28)$$

$$u_*\theta_* = \frac{a^2}{R} \ u \ \Delta\theta \ F_h \ (z/z_0 \ , \ R_iB) \qquad (8.29)$$

where

$$a^2 = k^2 \ / \ (\ln \ z/z_0)^2 \qquad (8.30)$$

is the drag coefficient in neutral conditions. The functions a^2F_m and a^2F_h have been computed numerically (Louis, 1979). These curves were then fitted by analytical formulae to avoid iterative computation at each time step during model integrations. Several considerations (see Louis, 1979) have lead to the following forms:

$$F = 1 - \frac{b \ R_i B}{1 \ + \ c |R_iB|^{1/2}} \qquad\qquad \text{unstable case} \qquad (8.31)$$

and

$$F = \frac{1}{(1 \ + \ b' \ R_iB)^2} \qquad\qquad \text{stable case} \qquad (8.32)$$

The values of b, b', and c were rather uncertain because of the large scatter in the observations. For numerical stability, the first derivative of the function F was constrained to be continuous between the stable and unstable regimes, which lead to $b = 2b' = 9.4$. The constant c was obtained by dimensional analysis in the free convection limit (Louis, 1979). At this point, it was assumed that for free convection the temperature gradient is a function of z, $g / \bar{\theta}$ and the heat flux only. Louis (1979) then introduced a temperature scale defined by

$$\theta_* = \left[\frac{\overline{w'\theta'^2}\,\bar{\theta}}{gz} \right]^{1/3} \tag{8.33}$$

and reformulated the heat flux profile as

$$\frac{z}{\theta_*} \frac{\partial \theta}{\partial z} = c \tag{8.34}$$

where c is a constant. Substituting for θ_* from (8.34) and integrating (8.33) from z_0 to z in the constant flux layer gives

$$\overline{w'\theta'} = c' \left[\frac{gz}{\theta} \right]^{1/2} \frac{\Delta\theta^{3/2}}{[(z/z_0)^{1/3}-1]^{3/2}} \tag{8.35}$$

where c' is a constant. Using (8.31) and considering the limit of RiB when $u \to 0$, Louis (1979) obtained

$$\overline{w'\theta'} = -u_*\theta_* = \frac{a^2b}{cR} \left[\frac{gz}{\theta} \right]^{1/2} \Delta\theta^{3/2} \tag{8.36}$$

Assuming z is much larger than z_0, (8.35) and (8.36) give

$$c = C^* a^2 b\, (z/z_0)^{\frac{1}{2}} \tag{8.37}$$

For a best fit of the computed F_m and F_h, Louis (1979) found $C^* = 7.4$ for momentum and $C^* = 5.3$ for heat and moisture fluxes. Finally the turbulent exchange coefficients for momentum, heat and moisture are obtained respectively, as

$$C_d = \frac{k^2}{(\ln\, z/z_0)^2} F_M \qquad\qquad \text{for momentum} \tag{8.38}$$

$$C_\theta = C_q = \frac{-k^2}{0.74\,(\ln\,z/z_0)^2}\,F_H \qquad \text{for heat and moisture} \qquad (8.39)$$

The functions F_M and F_H are obtained from (8.31) and (8.32) as

a)　　　　　　　Stable case; $RiB \geq 0$

$$F_M = F_H = \frac{1}{1 + 4.7RiB} \qquad\qquad (8.40)$$

b)　　　　　　　Unstable case; $RiB < 0$

$$F_M = 1 - \frac{9.4\,RiB}{1 + C_M\,|RiB|^{\frac{1}{2}}} \qquad\qquad (8.41)$$

and

$$F_H = 1 - \frac{9.4\,RiB}{1 + C_H\,|RiB|^{\frac{1}{2}}} \qquad\qquad (8.42)$$

where C_M and C_H are computed using (8.37) with appropriate values of C^*.

The fluxes of momentum, heat and moisture are then calculated as

$$M = \rho\,C_d\,(\bar{u}_2 - \bar{u}_1)^2 \qquad\qquad (8.43)$$

$$H = \rho\,C_p\,C_\theta\,(\bar{u}_2 - \bar{u}_1)(\bar{\theta}_2 - \bar{\theta}_1) \qquad\qquad (8.44)$$

$$Q = \rho\,L\,C_q\,(\bar{u}_2 - \bar{u}_1)(\bar{q}_2 - \bar{q}_1) \qquad\qquad (8.45)$$

where ρ is the air density, L is the latent heat of evaporation and C_p is the specific heat of air at constant pressure. Q is referred to as the potential evaporation and is scaled by a ground wetness parameter to obtain the actual evaporation rate. This parameter constitutes a measure of soil moisture availability and ranges from zero for completely dry grounds to unity for oceans. Ground moisture availability is a parameter difficult to predict and is empirically determined in many numerical weather prediction models. An example of ground wetness field estimated from surface albedo is shown in Figure (8.2).

SOIL MOISTURE FIELD

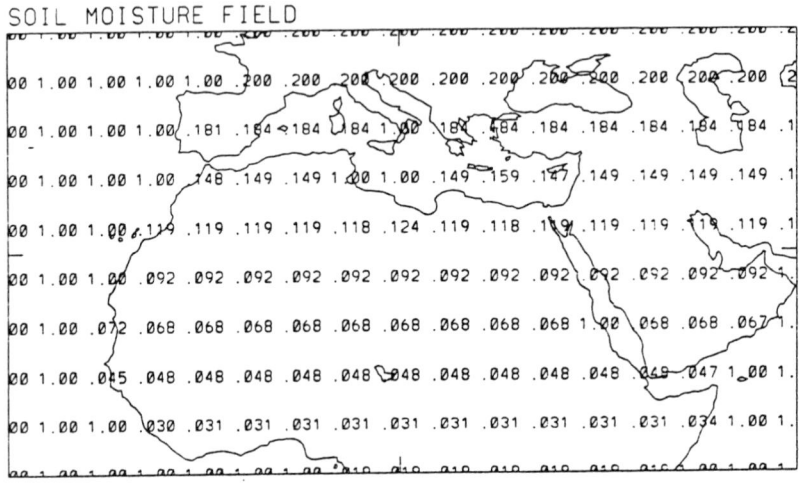

Figure 8.2: Ground wetness parameter as a function of surface
albedo (α).

$$GW = 0.85(1.0 - \exp(-200(0.25 - \alpha)^2)) \quad \text{for } \alpha < 0.25$$
$$GW = 0 \quad \text{for } \alpha \geq 0.25$$

Program (*SIMFLX*), provided in this section, gives a complete
illustration on the use of the similarity theory in surface fluxes
calculations. This simple column model uses input surface data for a
particular point ($I = 6$, $J = 7$) from the array defined in Fig. (8.2) and
makes use of subroutine *FLXSRF* to estimate the surface fluxes of heat,
moisture and momentum. Program (*SIMFLX*) can easily be extended to
compute surface fluxes over a larger domain provided the required data
are available.

program *SIMFLX*
```
c
c     This program estimates the surface fluxes of heat,
c     moisture and momentum using the similarity theory.
c     Surface fluxes calculations require too many surface
c     parameters which need to be defined at each grid point.
c     Because of the voluminous data required, this program
c     is designed to run over one grid point only. It is a
c     column model for surface flux calculations.
```

```
c     Although the surface fluxes are calculated using
c     only the anemometer level and surface variables, the
c     code presented here is set closely compatible to that
c     which would normally run in a Numerical Weather Prediction
c     model, where data comes into a vertical array on sigma
c     levels for each grid point (sigma = p/ps) where p is the
c     pressure at the given level and ps is the surface pressure.
c
         parameter ( nk = 12 )
         real t(nk),q(nk),sh(nk),s(nk)
         common/wet/eps,ae(2),be(2),ht(2),tc
         common/sfc/rib,ustr,ol,cd,ch,cq,z2,z1,sigz2,zv,zt
c
c     Parameters definition.
c
c     parsol : Ice absorption coefficient (over ocean)
c     cfonte : Snow melting coefficient (over land)
c     humsol : Soil moisture availability
c     albedo : Surface albedo
c     flxiri  : Outgoing longwave  radiation at surface
c     flxvis  : Incoming Shortwave radiation at surface
c     z0      : Roughness length
c     vent    : Wind speed at lowest model level
c     ts      : Surface temperature   (sst over oceans)
c     ps      : Surface pressure
c
c     Land Sea matrix
c
c     ntypes = -1  .... Land
c     ntypes =  0  .... Ocean
c     ntypes =  1  .... Land with snow
c     ntypes =  2  .... Ocean with ice
c
c     Define the sigma levels
c
         data s /.1,.2,.3,.4,.5,.6,.7,.8,.85,.9,.95,.99/
c
c     Define temperature and specific humidity profile for
c     the given grid point.
c
         data t /220.16,218.06,223.71,235.78,247.31,256.73,
      &          264.38,269.16,271.98,274.71,277.10,278.41/
c
```

```
          data q/ 4.0412000e-05,1.7426601e-05,2.6705300e-05
     &           ,9.5105803e-05,2.6963701e-04,6.8341498e-04
     &           ,1.6110500e-03,2.1570900e-03,2.3708199e-03
     &           ,2.5780201e-03,2.9228700e-03,3.2207901e-03/
c
c    Define surface parameters.
c
          data parsol,z0,ntypes,humsol/ 0.45,1.26,-1,0.15/
          data ts,ps,albedo      /278.96,88230.5,8.05e-02/
          data vent,flxvis,flxiri   /9.43,567.93,178.22/
c
c    Compute intermediate sigma levels.
c
          do 8100 k = 1,nk-1
 8100        sh(k)  = sqrt ( s(k)*s(k+1) )
          sh(nk) = sqrt ( s(nk) )
c
c    Define moisture constants.
c
          call WETCNS
c
c    Call surface fluxes routine.
c
          call flxsrf(flxqdm,flxcha,flxvap,ts,parsol,t(nk)
     &               ,flxiri,flxvis,vent,z0,ntypes,albedo
     &               ,humsol,q(nk),ps,s(nk),sh(nk))
c
c    Display Output surface fluxes.
c
          print *,flxqdm,flxcha,flxvap
          stop
          end
```

An algorithm showing the computation of the surface fluxes over a limited domain is also provided, (*SIMFLX_EXP*), but is not implemented because of the voluminous input data. This segment of code is important in the sense it shows the reader the different steps and their sequences for the computation of the surface fluxes within a numerical model. The code is extensively documented.

program *SIMFLX_EXP*

```
c
c       This program shows the calculation of the surface fluxes
c       for a limited domain. Calculations are presented as if
c       they were in a numerical model except that there is no time
c       integration.This case would correspond to fluxes computation
c       for one time step for all grid points.
c
c       Parameters definition:
c
c       l        : east-west dimension
c       m        : south-north dimension
c       nk       : number of vertical levels
c       slat     : southernmost latitude
c       wlon     : westernmost  longitude
c       hz       : time of the day in hours
c       day      : day of the month
c       month    : month of the year
c       dphi     : grid distance in the north-south direction
c       dlambda  : grid distance in the east-west direction
c       kount    : counter of the number of time steps
c       pas      : the time step in secondes.
c       stebol   : Stefan-Boltzman constant
c
c
c       Variables definitions
c
c       Input:
c
c       wt       : tridimensional temperature field
c       wq       : tridimensional specific humidity field
c       t        : temperature profile for the considered point
c       q        : specific humdity           //
c       s        : sigma levels
c       sh       : intermediate sigma levels
c       declsc   : sun declination angle
c       alb      : surface albedo
c       sfps     : surface pressure
c       tss      : surface temperature
c       ven      : surface wind
c       rlon     : longitude
c       alat     : latitude
c       slat     : southernmost latitude
```

```
c      wlong : westernmost longitude
c
c      Output:
c
c      warm   : heating rate
c      cool   : cooling rate
c      rtrter : vertical profile of terrestrial radiative flux
c      rtrsol : vertical profile of solar radiative flux
c      flxvis : flux of shortwave radiation at surface
c      flxiri : flux of infra-red radiation at surface
c      irtop  : infra-red radiation at top of atmosphere
c
c      common blocks wetcns and rrr define the moisture
c      constants and other variables used in the radiative
c      transfer calculations.
c      Calculations are assumed to be done over an array of
c      dimension (17,10) and over which all input data are
c      defined.
c
            parameter (l = 17,m = 10,nk = 12,nk2 = nk+2)
            real t(nk),q(nk),sh(nk),s(nk),declsc(2)
            real rtrter(nk),rtrsol(nk),irtop(l,m)
            real alb(l,m),sfps(l,m),tss(l,m),rlon(l,m)
            real ven(l,m),alat(l,m),cool(l,m,nk)
            real wt(l,m,nk),wq(l,m,nk),warm(l,m,nk)
            common/wetc/eps,ae(2),be(2),ht(2),tc
            common/rrr /fup(nk2,8),fdw(nk2,8),rw(nk2)
      &     ,fx(nk2),tl(nk2),aso(nk2,8),rso(nk2,8),albc(8),rsa,secz,
      &     coe(8),dtc(nk2,8),fnet(nk2,8),cld(nk2,8),rhx(nk),dp(nk),
      &     qz(nk),p(nk2),tz(nk),dpz(nk2),qs(nk),sso(8),dtw(nk2,8)
            dimension ntyp(l,m),pars(l,m),rlen(l,m),hums(l,m)
            real       flxq(l,m),flxc(l,m),flxv(l,m),topo(l,m)
            common /rhh  /rh(nk),dqq(nk)
            common /ipos /ihem,nbad,nbad1,nshal
            common /loc  /ic,jc
            common /extra/rib,ustr,ol,cd,ch,cq,z2,z1,sigz2,zv,zt
            common /density/rho
c
c      Input:
c
c      ntyp : land-sea matrix
c      pars : soil parameter
c      rlen : roughness lenth
```

```
c      hums : soil humdity
c      topo  : topography
c
c      Output:
c
c      flxq  : momentum flux
c      flxc  : heat flux
c      flxv  : moisture flux
c      swc and swr are logical switchs for shortwaves
c      and longwaves radiation calculations, respectively.
c
       logical swc,swr
       data swr ,swc /.true.,.true./
       data pi,dphi,dlambda /3.14159 ,5.5376   ,5.625 /
       data s /.1,.2,.3,.4,.5,.6,.7,.8,.85,.9,.95,.99/
       data pas,kount/1.,1./
       data stebol /5.6678e-8/
       data slat,wlong,hz,day,month/2.77,-28.13,12.,20,12/
       picon      = pi/180.
c
c      Open files for input data
c
       open (11,file='ts17.dat',status='old')
       open (12,file='ps17.dat',status='old')
       open (13,file='alb17.dat',status='old')
       open (14,file='vent17.dat',status='old')
       open (15,file='t17.dat',status='old')
       open (16,file='q17.dat',status='old')
       open (17,file='ntyp17.dat',status='old')
       open (18,file='pars17.dat',status='old')
       open (19,file='z017.dat',status='old')
       open (20,file='hums17.dat',status='old')
       open (21,file='xht17.dat ',status='old')
c
c      Open files for output
c
       open (22,file='warm.dat ',status='unknown ')
       open (23,file='cool.dat',status='unknown')
       open (24,file='surflx.dat',status='unknown')
c
 1000  format(6e13.6)
 1001  format(10(1x,17i2,/))
c
```

```
c      In normal model operations variables in files 11 to 21
c      are read in the initialization procedure and carried
c      by the model. In this example these variables are read in.
c
c      Read temperature and specific humdity for all levels
c
       do 8200 k  = 1, nk
           read (15,1000) ((wt(i,j,k),i=1,l),j=1,m)
           read (16,1001) ((wq(i,j,k),i=1,l),j=1,m)
 8200  continue
c
c      Read surface variables.
c
           read (11,100)((tss  (i,j),i=1,l),j=1,m)
           read (12,100)((sfps (i,j),i=1,l),j=1,m)
           read (13,100)((alb  (i,j),i=1,l),j=1,m)
           read (14,100)((ven  (i,j),i=1,l),j=1,m)
           read (17,101)((ntyp (i,j),i=1,l),j=1,m)
           read (18,100)((pars (i,j),i=1,l),j=1,m)
           read (19,100)((rlen (i,j),i=1,l),j=1,m)
           read (20,100)((hums (i,j),i=1,l),j=1,m)
           read (21,100)((topo (i,j),i=1,l),j=1,m)
c
c      Compute declination angle from day and month.
c
           call CONRAD (declsc,day,month)
c
c      Compute intermediate sigma levels (sh)
c
           do 8202 k = 1, nk-1
               sh(k)   = sqrt ( s(k)*s(k+1) )
 8202      continue
               sh(nk)  = sqrt ( s(nk) )
c
c      Main loop for all grid points.
c
           do 8204 j = 1, m
           do 8204 i = 1, l
c
c      Compute the position of the given grid point
c
           alat(i,j)   = (slat + (j-1)*dphi)*picon
           rlon(i,j)   = (wlon + (i-1)*dlambda)*picon
```

```
c
c     Prepare column arrays for the given grid point
c
      do 8206 k = 1, nk
         t(k)    = wt(i,j,k)
         q(k)    = wq(i,j,k)
8206  continue
c
c     Prepare surface parameters
c
      parsol  = pars (i,j)
      z0      = rlen (i,j)
      ntypes  = ntyp (i,j)
      humsol  = hums (i,j)
      xht1    = topo (i,j)
      ts      = tss  (i,j)
      ps      = sfps (i,j)
      rlong   = rlon (i,j)
      sinlat  = sin (alat(i,j))
      coslat  = cos (alat(i,j))
      albedo  = alb  (i,j)
      vent    = ven  (i,j)
      f       = 2.*7.29e-05*sinlat
      ic      = i
      jc      = j
c
c     Prepare moisture constants
c
      call WETCNS
c
c     Call radiation routine to obtain the infra-red and shortwave
c     radiative fluxes at the surface, these are required inputs
c     for surface fluxes calculation.
c
      call RAD (rtrsol,rtrter,flxvis,flxiri,ts,t,q,ps,sh,nk,
     1          rlong,sinlat,coslat,s,hz,albedo,vent,declsc,
     2          swr,swc,irtop,ic,jc)
c
c     Save output of warm and cool for the grid point
c
      do 8208 k = 1, nk
      warm(i,j,k) = rtrsol(k)
      cool(i,j,k) = rtrter(k)
```

```
 8208   continue
c
c     Call surface fluxes routine for the given grid point.
c
        call FLXSRF(flxqdm,flxcha,flxvap,ts,parsol,t(nk),flxiri,
     &             flxvis,vent,z0,ntypes,albedo,humsol,f,ch1,ch2,
     &             ch3,q(nk),ps,s(nk),pas,sh(nk),xht1,kount)
c
c     Save surface fluxes output for the given grid point
c
        flxq(i,j)   = flxqdm
        flxc(i,j)   = flxcha
        flxv(i,j)   = flxvap
c
c     Continue for the next point
c
 8204   continue
        stop
        end
```

4. Height of the Boundary Layer in an Unstable Situation

The vertical distribution of the surface fluxes within the lowest tropospheric levels requires an estimation of the height of the planetary boundary layer. The stable and unstable planetary boundary layers are treated differently, and the definition of the stability is based on the bulk Richardson's number.

Relevant references for this section may be found in Deardorff (1972), Smeda (1977) and the various reports on the planetary boundary layer published by the European Center for Medium Range Weather Forecast (ECMWF). In the unstable case, the Monin-Obukhov length, L, the bulk Richardson number, R_iB, and the modified bulk Richardson number are negative. The surface heat flux is upward, i.e. $-w'\theta' > 0$ and $\theta^* > 0$. Let h denote the height of the planetary boundary layer. Air is unstable below h and is stable above it. Across the interface $(z = h)$, two infinitesimal layers are defined. The upper and lower infinitesimal layers' potential temperatures are denoted by θ^+ and θ^-, respectively. Let $\Delta\theta = \theta^+ - \theta^-$ and require that $\dfrac{\partial \Delta\theta}{\partial t} = 0$. This assures the finiteness of the interfacial gradient, i.e.,

$$\frac{\partial \theta^+}{\partial t} = \frac{\partial \theta^-}{\partial t} \qquad (8.46)$$

In the well-mixed layer the governing equation for the change of potential temperature is

$$\frac{\partial \theta^-}{\partial t} = -\frac{\partial}{\partial z} \overline{w'\theta'} \qquad (8.47)$$

It is assumed that the flux $\overline{w'\theta'}$ decreases linearly away from the surface, z_1, and vanishes at $z = h$. Thus,

$$\overline{w'\theta'}(z) = \overline{w'\theta'})z_1 \, (h - z/h) \qquad (8.48)$$

$$\frac{\partial}{\partial z} \frac{\overline{w'\theta'}}{} = \frac{-\overline{w'\theta'})z_1}{h}$$

In the stable layer ($z \geq h$) the change of θ^+ arises due to the large scale convection and the changes in the interfacial height h. If w^+ is the large scale vertical velocity, then

$$\frac{\partial \theta^+}{\partial t} = -\,(w^+ \frac{\partial \theta^+}{\partial z} - \frac{dh}{dt} \frac{\partial \theta^+)}{\partial z}) \qquad (8.49)$$

where $\dfrac{dh}{dt}$ is the rate of change of the height of the interface. Equating (8.47) and (8.49) and using (8.48) yields

$$\frac{dh}{dt} - w^+ = \frac{\overline{w'\theta'})z_1}{h \, \dfrac{\partial \theta^+}{\partial z}} \qquad (8.50)$$

It should be noted that the heat flux at the lower boundary is obtained from the surface flux calculations for the unstable stratification (this was the similarity solution discussed earlier), $\dfrac{\partial \theta^+}{\partial z}$ is the stability above the planetary boundary layer and is a known quantity since it can be determined from large scale data. W^+ is the large scale vertical velocity at the top of the planetary boundary layer and is again defined over the large scale grid. Thus, the only unknown in the above equation is h as a function of time t, for a given height z. Since the horizontal advection of h is small, $\frac{dh}{dt}$ can be replaced by its local variation.

$$\frac{\partial h}{\partial t} = w^+ + \frac{\overline{(w'\theta')}z_1}{h\,\frac{\partial\theta^+}{\partial z}} \tag{8.51}$$

It can be shown that (8.51) has a solution of the form

$$h = tw^+ + \frac{\overline{(w'\theta')}z_1}{w^+\,\frac{\partial\theta^+}{\partial z}}\ln h + \frac{\overline{(w'\theta')}z_1}{w^+\,\frac{\partial\theta^+}{\partial z}} + Cw^+ \tag{8.52}$$

where C is an integration constant. However, this solution is transcendental in h and has a limited value in this form. Since the effect of the large scale vertical velocity is smaller than that of the vertical heat flux in unstable conditions, a simplified solution for (8.51) may be obtained as

$$h^2 = \frac{\overline{(w'\theta')}z_1}{\frac{\partial\theta^+}{\partial z}}\,t + C \tag{8.53}$$

or

$$h = \sqrt{2at + h_0^2} \tag{8.54}$$

where $a = \overline{(w'\theta')}z_1 / (\partial\theta^+/\partial z)$ and h_0 is an initial height of the planetary boundary layer. Thus, h increases slowly as the heat flux remains upward, and larger stability of the air above level h slows its growth. Separating the growth of h due to large scale vertical velocity from that due to the eddy heat flux, an approximate solution may be written in the form,

$$h = wt + \sqrt{2at + h_0^2} \qquad (8.55)$$

In practical application this requires a knowledge of $h_0(x,y)$ at each time step.

5. Height of the Planetary Boundary Layer in a Stable Situation

The formulation of the stable case is a difficult problem since the rate of change of the height of the planetary boundary layer is not easily definable. No theories on the behavior of the so-called nocturnal boundary layer exist for large scale numerical weather prediction. Although several rate equations have appeared in recent literature, their applications have been of limited value. According to Deardorff (1972) the time rate of change of the height of the planetary boundary layer may be expressed by

$$\frac{\partial h}{\partial t} = \frac{2u_*^2}{\beta h^2 \Upsilon^2 + 7u_*^2} \left[1 - 10 \frac{fh^2}{u_*} \right] - Cu_* \frac{h}{L} \qquad (8.56)$$

where β is the buoyancy parameter and L denotes the Monin Obukhov length. The parameter f is the Coriolis parameter and Υ represents the stability $\frac{\partial \theta^+}{\partial z}$ of a layer immediately above the planetary boundary layer h. C is an empirical constant whose value is around 5×10^{-3}. That is a best fit value based on **WANGARA** field experiment. The principal idea here is that the growth of the planetary boundary layer depends on the stress induced by the near surface wind. The last term provides a somewhat reasonable transition from an unstable to a stable planetary boundary layer.

6. Vertical Distribution of Fluxes

At the lower boundary of the atmosphere, the vertical diffusion of momentum, heat and moisture is described using the so-called "K-theory". This is generally expressed as

$$\frac{\partial \tau}{\partial t} = \frac{1}{\rho} \frac{\partial}{\partial z} \left[\rho k \frac{\partial \tau}{\partial z} \right] \tag{8.57}$$

where τ is a generalized variable which may represent u, v, θ or q, and k represents the eddy diffusion coefficient. The diffusion coefficients are a function of the vertical wind shear, the mixing length and the bulk Richardson number RiB,

$$k_h = k_q = \ell^2 \left| \frac{\partial \vec{v}}{\partial z} \right| F_1 \ (RiB) \qquad \text{for heat and moisture}$$

$$\tag{8.58}$$

$$k_m = \ell^2 \left| \frac{\partial \vec{v}}{\partial z} \right| F_2 \ (RiB) \qquad \text{for momentum}$$

Blackadar (1962) suggested a relation for the estimation of the mixing length as

$$\ell = \frac{kz}{(1 \ + \ kz/\lambda)} \tag{8.59}$$

where k is the Von Kármán constant and λ is a maximum mixing length given by

$$\lambda = 150 \text{ m} \qquad \text{for momentum,}$$
$$\lambda = 450 \text{ m} \qquad \text{for heat and moisture.}$$

The functional relations for the stability of the planetary boundary layer are expressed as

$$F_1 = F_2 = \frac{1}{(1 \ + \ 5 \ RiB)^2} \qquad \text{for } RiB \geq 0 \tag{8.60}$$

and

$$F_1 = 1 - \frac{8RiB}{1 \ + \ 1.286 \left| RiB \right|^{\frac{1}{2}}} \qquad \text{for } RiB < 0 \tag{8.61}$$

$$F_2 = 1 - \frac{8RiB}{1 \ + \ 1.746 \left| RiB \right|^{\frac{1}{2}}} \qquad \text{for } RiB < 0 \tag{8.62}$$

The bulk Richardson number is defined as

$$RiB = \frac{g}{\theta}\frac{\partial \bar{\theta}}{\partial z} \left/ \left| \frac{\partial \vec{v}}{\partial z} \right|^2 \right. \tag{8.63}$$

The effect of moisture is introduced by the use of virtual potential temperature. It is obvious that the solution of the diffusion equation (8.57) requires boundary conditions. Generally the fluxes are set to zero at the top boundary and equated to the surface fluxes at the lower boundary. It is important to note that small vertical wind shear tends to produce large Richardson numbers. Consequently the eddy diffusivity coefficients approach zero in the stable case leading to unrealistically large flux convergence within the layer. This problem is avoided by imposing $\left| \frac{\partial \vec{v}}{\partial z} \right| \geq 1$ ms^{-1}. Moreover, the diffusion equation (8.57) simplicitly invokes a time scale of the order of $\frac{1}{T} = \frac{k}{\Delta z^2}$. In practice, T is set to a minimum of one day. In case $\frac{\Delta z^2}{k}$ does not satisfy this condition, the coefficient k is recalculated as $k = \frac{\Delta z^2}{T}$ to ensure the diffusion acts on a time scale of one day or more.

The vertical distribution of the surface fluxes also depends on the stability of the planetary boundary layer as defined by the sign of the Monin Obukhov length L. In case of unstable conditions (L < 0), the planetary boundary layer is well mixed and the vertical profiles of u, v, θ and q are almost invariant with height. The maintenance of this structure requires linear fluxes profiles so that the fluxes divergence remains linear throughout the depth of the mixed layer. The linear flux profile is then described by the relation,

$$\frac{F}{F_s} = 1 - \frac{z}{H_p} \tag{8.64}$$

where F and F_s are the fluxes at level z and at the surface, respectively. H_p is the height of the planetary boundary layer. The boundary conditions are expressed as

$$\begin{cases} F = 0 & \text{at } z = H_p \\ \\ F = F_s & \text{at } z = 0 \end{cases} \tag{8.65}$$

It is important to note that these boundary conditions do not allow for counter-gradient heat flux at the top of the planetary boundary layer, which may arise as a result of warm entrainment from above the inversion (Lenschow, 1970). The flux convergence for the linear profile is then obtained as

$$\frac{1}{\rho}\frac{\partial F}{\partial z} = - \frac{F_s}{\rho_s H_p} \tag{8.66}$$

where ρ_s is the air density at the surface. The diffusion equation may then be expressed by

$$\frac{\partial \tau}{\partial t} = - \frac{F_s}{\rho_s H_p} \tag{8.67}$$

where

$$\tau = u,v \qquad\qquad \text{for momentum}$$
$$\tag{8.68}$$
$$\tau = \theta \qquad\qquad \text{for heat or moisture}$$

Even though simple treatment of the fluxes distribution is obtained for the unstable surface conditions through the mixed layer theory, this assumption does not hold for the stable case. For this case a quadratic representation of the surface fluxes is suggested,

$$F = az^2 + bz + c \tag{8.69}$$

This formulation generates a large flux convergence at the top of the planetary boundary layer. Using the same boundary condition as in (8.65) and constraining a zero flux convergence at the top of the planetary boundary layer, i.e.,

$$\frac{\partial F}{\partial z} = 0 \qquad \text{at } z = H_p \tag{8.70}$$

one obtains

$$a = - \frac{F_s}{H_p^2} \quad , \quad b = \frac{2F_s}{H_p} \quad , \quad c = - F_s \tag{8.71}$$

Following the same steps as for the unstable case, the diffusion equation expressed as

$$\frac{\partial \tau}{\partial t} = \frac{1}{\rho}\frac{\partial F}{\partial z} = \frac{2F_s}{\rho_s H_p}\left[1 - \frac{z}{H_p} \right] \tag{8.72}$$

will generate linear profiles of stress, heating or moistening for momentum, sensible or latent heat, respectively.

9

Radiative Transfers

Calculations of the radiative transfers in the atmosphere represent one of the most important points in numerical weather prediction. The radiative heating constitutes an essential element for the maintenance of the global atmospheric heat budget and contributes an important part to the total energy driving the earth climate. The solar energy at the top of the atmosphere undergoes many alterations before it finally reaches the earth's surface. Part of this energy is reflected at the top of the atmosphere defining the planetary albedo while the other part crosses the atmosphere. The part that passes through the atmosphere undergoes scattering, absorption and reflection by atmospheric constituents and only a fraction of it reaches the ground. Part of this solar radiance is reflected on the ground defining the surface albedo. The remaining part, together with the net downward longwave irradiance emanating from the atmosphere, constitute the net radiation absorbed at the surface. The bulk of the absorbed energy is redistributed to the atmosphere via eddy fluxes of sensible and latent heat, while a small fraction is exchanged with the ground superficial layers. With the development of numerical weather prediction models, many radiative processes parameterization schemes with various degrees of complexity have been developed to describe the effect of radiative energy on the earth-atmosphere system. It would, therefore, be lengthy to describe all of them. A detailed description of one method that has been in use at Florida State University is presented in this book. This method uses an emissivity tabulation for the longwave or infrared radiation and an absorptivity function to define the shortwave radiative transfer. The scheme uses a simple surface energy balance to estimate the ground temperature and the sensible and latent heat fluxes over land points. The model also includes the calculation of the solar zenith angle which allows for diurnal and seasonal cycles and a detailed description of the cloud cover based on threshold relative humidity.

1. Longwave Radiation

The integration of the transfer equations over the entire infrared spectrum and optical depth is a complex and time consuming process.

Simplified schemes adapted to present computers' capability are then developed for use in numerical weather prediction. The performance of the emissivity technique has been reviewed in many studies (e.g., Rodgers, 1967; Fels and Schwarzkopf, 1981) and was found adequate for most numerical weather prediction models. The method consists of expressing the emissivity as a function of the path length which in turn depends on temperature, pressure and relative humidity. The basic emissivity tabulation is a contribution from Staley and Jurice (1970). The actual value of emissivity is then interpolated given the predicted path length.

Although carbon dioxide and ozone constitute important absorbents of longwave infrared radiation, only absorption and emission by water vapor is considered in this simplified model. It should be noted however that water vapor is a major atmospheric constituent and has a strong absorption in the vibrational and rotational bands around 6 and 20 μm, respectively.

Additional assumptions made in simplifying the longwave radiative transfer equations may be summarized as follows:

- Clouds are considered as an infinite isothermal atmosphere which radiates as a blackbody.
- The earth's surface is considered as blackbody.
- The atmosphere is considered as a stratification of horizontally homogeneous plane-parallel layers.
- Air molecules scattering is neglected and isotropic atmosphere is assumed.

For a cloud free atmosphere, the upward flux of longwave radiation at a reference level i may be divided into two parts, (Figure 9.1).

The first part emanates from the ground surface at temperature T_g and reaches the reference level. This is written as

$$F_g{\uparrow} = \sigma T_g{}^4 (1 - \varepsilon [w_b - w_i]) \tag{9.1}$$

The second part is emitted by the layer comprised between the ground and the reference level and is expressed as

$$F{\uparrow} = \int_{w_i}^{w_b} \sigma T^4 \frac{\partial \varepsilon(w - w_i)}{\partial w} \, dw \tag{9.2}$$

where σ is the Stefan-Boltzmann constant, w_b is the path length at the

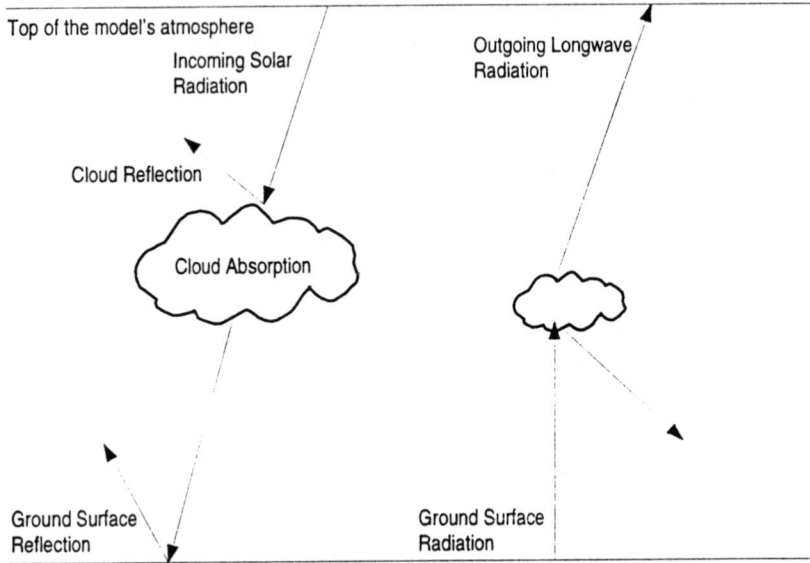

Figure 9.1: Schematic representation of radiative fluxes.

ground and ε is the emissivity and is function of the path length. The contribution from the region above the model atmosphere $(0 < p < P_T)$ is usually constructed from a standard atmospheric sounding. The optical depth at pressure p is given by

$$w(p) = \frac{1}{g} \int_{P_t}^{P_b} q\left[\frac{P}{P_0}\right]^{0.85} \left[\frac{T_0}{T}\right]^{0.5} dp + w_t \qquad (9.3)$$

where q is the specific humidity and w_t represents the optical path above the model atmosphere. It is suggested that the evaluation of the optical path takes into consideration the reduction of pressure and the temperature correction expressed by the terms $\left[\dfrac{P}{P_0}\right]^{0.85}$ and $\left[\dfrac{T_0}{T}\right]^{0.5}$, respectively. The specification of w_t assumes an isothermal stratosphere and a frost point of $210°K$. This constitutes a measure of the water vapor in the stratosphere.

The total upward infrared radiation flux at the reference level is then given by

$$F_i\uparrow = \sigma T_g^4 (1 - \varepsilon [w_b - w_i]) + \int_{w_i}^{w_b} \sigma T^4 \frac{\partial \varepsilon(w - w_i)}{\partial w} dw \quad (9.4)$$

The flux emissivity $\varepsilon(w - w_i)$ is defined by

$$\varepsilon(w - w_i) = \frac{\pi}{\sigma T^4} \int_0^\infty B_v(T) (1 - \tau_v [w - w_i]) dv \quad (9.5)$$

where $\tau_v (w - w_i)$ is the transmission function evaluated for a path length $(w - w_i)$, and $B_v(T)$ is the Planck blackbody radiance at frequency v and temperature T.

The downward flux of longwave radiation reaching the reference level in the cloud free case is given by

$$F_i\downarrow = - \int_0^{w_i} \sigma T^4 \frac{\partial \varepsilon(w_i - w)}{\partial w} dw + I_t \quad (9.6)$$

where I_t is the incoming longwave irradiance at the top of the model atmosphere.

In case a cloud layer exists above the reference level, the downward longwave flux is altered to yield

$$F_i\downarrow = \sigma T_{cb}^4 (1 - \varepsilon [w_i - w_{cb}]) - \int_{w_{cb}}^{w_i} \sigma T^4 \frac{\partial \varepsilon(w_i - w)}{\partial w} dw \quad (9.7)$$

where w_{cb} and T_{cb} represent the optical path and temperature at the cloud base, respectively. If there were a cloud layer below the reference level, the upward flux would then be

$$F_i\uparrow = \sigma T_{ct}^4 (1 - \varepsilon[w_{ct} - w_i]) - \int_{w_i}^{w_{ct}} \sigma T^4 \frac{\partial \varepsilon(w - w_i)}{\partial w} dw \quad (9.8)$$

where the subscript, "ct", stands for cloud top. In the case of a multiple cloud layers, a simple extension of these principles is applied.

The net longwave radiative flux at the reference level is then obtained as

$$F_i = F_i{\downarrow} - F_i{\uparrow} \tag{9.9}$$

and the warming (cooling) due to longwave radiation is determined by the convergence (divergence) of the net flux as

$$\left[\frac{\partial T}{\partial t} \right]_{IR} = \frac{g}{c_p} \frac{\partial F}{\partial p} \tag{9.10}$$

2. Shortwave Radiation

Solar radiation incident at the top of the atmosphere is generally separated into absorbed and scattered part (Joseph, 1966; Katayama, 1972). The depletion of shortwave radiation is caused by water vapor absorption and Rayleigh scattering by aerosols. The treatment of aerosols is too complex and is not considered in this model. Following Joseph and Katayama, Rayleigh scattering is significant in the wavelength domain less than 0.9 μm, where absorption by water is negligible. In the near infrared region with wavelengths greater than 0.9 μm the reverse process is observed. The absorbed and scattered parts are respectively expressed as

$$S_a = 0.349 \ S_o \cos \xi \tag{9.11}$$
$$S_s = 0.651 \ S_o \cos \xi \tag{9.12}$$

where S_o is the solar constant and is taken equal to 1.96 cal cm^{-2}min^{-1} or 1367 Wm^{-2}, ξ is the zenith angle and $S_o\cos\xi$ represents the incident insolation reaching the top of the atmosphere. Only the absorbed part of radiation contributes to warming of the atmosphere, but both components are absorbed and reflected at the ground surface. The zenith angle varies as a function of the position and time, and is expressed as

$$\cos \xi = \sin(\psi) \ \sin(\delta) + \cos(\psi) \ \cos(\delta) \ \cos(h) \tag{9.13}$$

where ψ is the latitude, δ is the solar declination angle and h is the hour angle measured from the local noon. The absorption by water vapor is approximated by an empirical relation (Arakawa, 1971).

$$\text{Absorbed Irradiance} = 0.189 \ (\text{w sec } \xi)^{0.303} \ \text{ly/min} \tag{9.14}$$

For a mean incident insolation ($S_0 \cos \xi$) of 2 ly/min, the absorptivity due to water vapor with respect to the absorbed part of insolation becomes

$$A(w) = \frac{0.189 \; [w \; \sec\xi]^{0.303}}{0.349 \; S_0 \; \cos\xi} \tag{9.15}$$

or

$$A(w) = 0.271 \; [w \; \sec \xi]^{0.303} \tag{9.16}$$

At the reference level i, the direct shortwave radiation may be expressed as

$$S_a = (1 - A[w_i \; \sec \xi]) \tag{9.17}$$

After reflections at the earth's surface or at cloud top and transmission through a cloud layer, the direct solar rays are isotropically diffused. Downward reflection at the cloud base is neglected. Hence, there is no multiple reflection between the cloud base and the earth's surface. Because cloud albedo is greater than 0.5, the absorption of reflected solar energy is important near the cloud top atmosphere, except in the case of high clouds where most of the reflected solar radiation is returned to space due to the small amount of water vapor in the upper atmosphere. In general, the diffuse radiation experiences a longer path length compared to direct solar radiation. Joseph (1966) suggested that the absorptivity for diffuse radiation be expressed as $A(1.66[w])$ instead of $A(w)$, where the factor 1.66 accounts for the increased path length. The diffuse radiation that reaches the reference level from the earth's surface is then expressed as

$$S_a\{1 - A[w_0 \cos \xi]\} \; \alpha_s \; \{1 - A[1.66(w_0 - w_i)]\} \tag{9.18}$$

and the total downward flux of solar radiation may be given by

$$S^i = S_a\{1 - A[w_i \; \xi]\} - S_a\{1 - A[w_0 \sec \xi]\} \; \alpha_s \; \{1 - A[1.66(w_0 - w_i)]\} \tag{9.19}$$

If a cloud layer exists just below the reference level, the diffuse radiance emanating upward at the cloud level is defined as

$$S_a(1 - A[w_{ct} \; \sec \xi]) \; \alpha_c \tag{9.20}$$

where w_{ct} and α_c are the optical path and albedo at the cloud top, respectively. Part of this energy is absorbed in the layer between the cloud top and the reference level. The ascending flux of diffuse radiation reaching the reference level is written as

$$S_a(1 - A[w_{ct} \sec \xi]) \, \alpha_c \, (1 - A[1.66(w_{ct} - w_i)]) \tag{9.21}$$

Therefore, in the presence of a cloud layer below the reference level, the net flux of solar radiation takes the following form:

$$\overset{i}{S_a} = S_a(1 - A[w_i \sec \xi]) - S_a(1 - A[w_{ct} \sec \xi]) \, \alpha_c \, (1$$

$$- A[1.66(w_{ct} - w_i)]) \tag{9.22}$$

The next step consists of examining the amount of shortwave radiation passing through a cloud layer. The definition of the absorptivity of the cloud is required for this purpose. Katayama (1972) defined the cloud's absorptivity function as $A[w_c^*]$, where w_c^* is the augmented path length which takes into account the equivalent amount of water vapor within the cloud. If S_a is the absorbed shortwave radiation reaching the top of the atmosphere, $S_a(1 - A[w_{ct} \sec \xi])$ would be the part reaching the cloud top. The amount reaching the cloud bottom can be expressed as

$$S_a(1 - A[w_{ct} \sec \xi])(1 - \alpha_c)(1 - A[w_c^*]) \tag{9.23}$$

The determination of the total flux of net shortwave radiation below a single cloud requires the consideration of the upward diffuse shortwave rising from the ground surface. This latter is carried out just as for the cloud free case. Finally the total downward flux of absorbed shortwave radiation at a reference level below a single cloud layer is expressed as

$$\overset{i}{S_a} = S_a(1 - A[w_{ct} \sec \xi])(1 - \alpha_c)\left\{ (1 - A[w_c^* + 1.66(w_i - w_{cb})]) - (1 \right.$$

$$\left. - A[w_c^* + 1.66(w_o - w_{cb})]) \, \alpha_s \, (1 - A[1.66(w_o - w_i)]) \right\} \tag{9.24}$$

where w_c^* represents the equivalent path length of the cloud and w_0 and w_{cb} are the optical depths at the ground and cloud base, respectively. This process can be extended for the treatment of multiple cloud layers.

In general, the atmospheric heating rate due to scattered shortwave radiation is small, but it plays a non-negligible role in the earth's surface heat balance. The theory of scattering is too involved and is beyond the scope of this workbook. Empirical relations (Chang, 1978) are usually used to define the scattered shortwave radiation at a reference level as

$$\overset{i}{S_s} = S_s (1 - \alpha_0)(1 - \alpha_0 \alpha_s) \tag{9.25}$$

where S_s is the scattered radiation at the atmosphere's top, and α_s is the earth's surface albedo. The albedo of the atmosphere is defined as a function of the surface pressure p_s and the zenith angle ξ and is expressed as

$$\alpha_0 = 0.085 - 0.245 \log(p_s / p_0 \cos \xi) \tag{9.26}$$

where $p_0 = 1000$ mb.

3. Cloud Specification

Clouds are the most important modulators of radiation in the atmosphere. Their presence has a dual effect of decreasing the incoming shortwave radiation reaching the ground surface and reducing the outgoing longwave radiation. The inclusion of cloud effects in radiative transfer calculations appears therefore essential.

Threshold relative humidity values following Slingo (1987) are used to characterize the presence of low (c_ℓ), middle (c_m) and high (c_h) clouds. These values are defined to be 66%, 50% and 40%, respectively, and are representative of mean tropical tropospheric situations. The fractional cloud in each category is described as

$$c_\ell = (\overline{Rh} - Rh_{c_\ell}) / (1 - Rh_{c_\ell}) \tag{9.27}$$

$$c_m = (\overline{Rh} - Rh_{cm}) / (1 - Rh_{cm}) \tag{9.28}$$

$$c_h = (\overline{Rh} - Rh_{ch}) / (1 - Rh_{ch}) \tag{9.29}$$

where $\overline{Rh}$ is the mean relative humidity of the layer and Rh_c is the critical value for a particular cloud type. It should be noted that all these three parameters are confined between 0 and 1 inclusive. Eight cloud configurations are described in the model. These are

1. Clear
2. Low cloud
3. Middle cloud
4. High cloud
5. Low and middle clouds
6. Low and high clouds
7. Middle and high clouds
8. Low, middle and high clouds

Different weights are assigned to the radiative fluxes depending on the cloud amount within a grid square area. These weights are defined as

$$c_1 = (1 - c_\ell)(1 - c_h)(1 - c_m) \qquad\qquad c_5 = c_\ell c_m (1 - c_h)$$

$$c_2 = c_\ell (1 - c_m)(1 - c_h) \qquad\qquad c_6 = c_\ell (1 - c_m) c_h$$

$$\qquad\qquad\qquad\qquad\qquad\qquad\qquad\qquad\qquad\qquad (9.30)$$

$$c_3 = (1 - c_\ell) c_m (1 - c_h) \qquad\qquad c_7 = (1 - c_\ell) c_m c_h$$

$$c_4 = (1 - c_\ell)(1 - c_m) c_h \qquad\qquad c_8 = c_\ell c_m c_h$$

It should be noted that this is not a unique method for cloud specification in radiative transfer calculations and that the threshold relative humidity values may be adjusted to account for regional differences.

4. Radiative Heat Balance at the Earth's Surface

In this simplified version of the radiative transfer model, the ground temperature is solved using a heat balance condition at the surface level. Although large diurnal changes may occur in the ground temperature evolution, the balance condition is acceptable to some extent, since the ground heat capacity is small. Under this assumption, the net absorbed radiative energy is set to balance the surface energy fluxes. It should be noted, however, that this assumption presents a serious shortcoming in the sense that it does not allow for heat flux exchanges through the earth's surface. The simplified balance is expressed as

$$R_\ell + R_s + H_s + H_\ell = 0 \qquad\qquad (9.31)$$

where R_ℓ is the net longwave radiation at the surface and has two components such as

$$R_\ell = F_\ell\!\downarrow - \sigma T_g^4 \tag{9.32}$$

$F_\ell\!\downarrow$ denotes the downward flux of longwave radiation, and σT_g^4 represents the upward flux of longwave radiation originating from the earth's surface. The second term in (9.31) is the net shortwave radiation reaching the ground surface, and it includes both the scattered and absorbed part of the solar radiation. It is expressed as

$$R_s = F_s\,(1 - \alpha_s) \tag{9.33}$$

where α_s is the ground surface albedo and F_s is the total downward shortwave flux reaching the ground. Finally, H_s and H_ℓ denote respectively the sensible and latent heat fluxes at the surface and are determined using similarity theory as described in Chapter 8. The surface energy balance (9.31) is coupled to the similarity theory and is solved for the ground temperature using the Newton-Raphson iterative algorithm. This coupling is necessary as the sensible and latent heat fluxes are ground temperature dependent. Over oceans the sea surface temperature is prescribed. Among the important parameters required for the solution of the surface heat balance are the surface albedo and the ground wetness parameters. Although it is recognized that these variables depend on the soil characteristics and vegetation, they are specified as time invariants in this simplified model. For short term numerical weather prediction problems, the use of seasonal albedo tabulations such as those prepared by Posey and Clapp (1964) and Kondratyev (1972) appears adequate. For climate prediction, however, monthly mean albedo values are recommended. The ground wetness parameter determines the ratio of effective evaporation to the potential evaporation and is expressed as

$$GW = E\,/\,E_p \tag{9.34}$$

While for short-range prediction (2 to 3 days) this parameter may, to some extent, be kept constant, it is imperative that it be time dependent in long range forecasts. The ground wetness parameter (soil moisture availability) represents a central modulator in the surface fluxes distribution (Bounoua and Krishnamurti, 1993). In this version of the model, the soil moisture availability is expressed empirically as a function of albedo as shown in Figure (8.2).

5. The Code

A detailed radiative transfer model based on the so-called emissivity method is presented in this section (*RADIA*). The model computes separately the longwave and shortwave radiations using subroutines *RLW* and *SLR*, respectively. The declination angle is calculated in subroutine *CONRAD* and the clouds specification is done in subprogram *RAD*. Because they are time consuming, radiative transfer calculations are not invoked at each time step during numerical weather prediction simulations. An update of the radiative transfer calculations at a 3 hour interval seems appropriate for most numerical weather prediction applications. A simple driver showing the use of this model for one atmospheric column is provided. Outputs illustrating the results from this column model are presented in Figure (9.2).

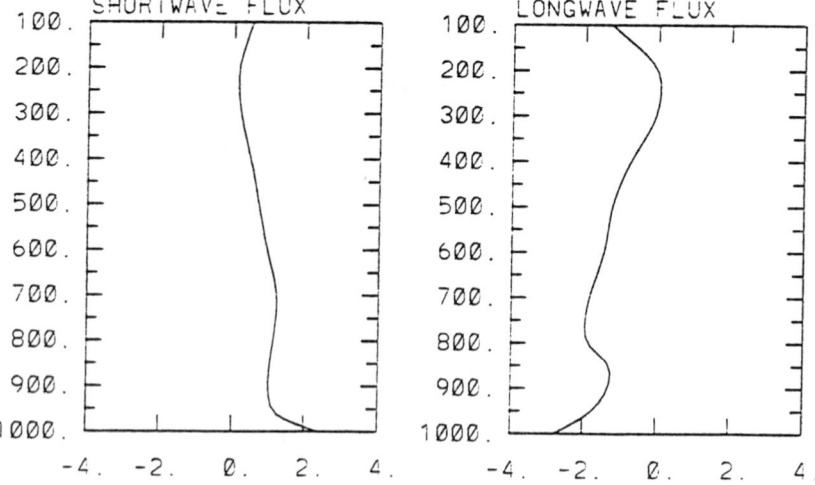

Figure 9.2: Vertical profiles of shortwave and longwave radiative fluxes program *RADIA*.

```
c
c      This program estimates the longwave and shotwave
c      radiative fluxes using the emissivity method.
c      CAUTION : On some machines this program needs
c      to be compiled with an option that initialize all
c      unitialized variables to zero.For most compilers
c      this procedure is the default.
c
       parameter ( nk =12)
       parameter ( hz = 12.,day = 20,month = 12)
       parameter (ilevv = 12,ilev2v = ilevv+2)
       real irtop
       dimension t(nk),q(nk),sh(nk),s(nk),declsc(2)
       dimension rtrter(nk),rtrsol(nk)
       common/wetc/eps,ae(2),be(2),ht(2),tc
       common/rrr/ fup(ilev2v,8),fdw(ilev2v,8),rw(ilev2v)
     & ,fx(ilev2v),tl(ilev2v),aso(ilev2v,8),rso(ilev2v,8)
     & ,albc(8),rsa,secz,coe(8),dtc(ilev2v,8),fnet(ilev2v,8)
     & ,cld(ilev2v,8),rhx(ilevv),dp(ilevv),qz(ilevv),p(ilev2v)
     & ,tz(ilevv),dpz(ilev2v),qs(ilevv),sso(8),dtw(ilev2v,8)
c
c      Define the logical switches for radiation calls.
c
       logical swc,swr
       data swr,swc /.true.,.true./
c
c      Define the grid distance in the x and y directions
c      and some other required constants.
c
       data dphi,dlambda,stebol /5.5376,5.625,5.6678e-8 /
c
c      Calculation are done on sigma levels.
c      sigma = p/ps.
c
       data s /.1,.2,.3,.4,.5,.6,.7,.8,.85,.9,.95,.99/
c
       pi         = 3.14159
       picon      = pi/180.
c
c      Open output file
c
       open (20,file='radia.out',status='unknown')
c
```

```
c     Define temperature ,specific humidity for the
c     vertical column.
c
      data t/220.16,218.06,223.71,235.78,247.31,256.73,
     &      264.38,269.16,271.98,274.71,277.10,278.41/
c
      data q/ 4.0412000e-05,1.7426601e-05,2.6705300e-05
     &       ,9.5105803e-05,2.6963701e-04,6.8341498e-04
     &       ,1.6110500e-03,2.1570900e-03,2.3708199e-03
     &       ,2.5780201e-03,2.9228700e-03,3.2207901e-03/
c
c     Define surface parameters. These variables are explained
c     in SIMFLX.
c
      data ts,ps /278.96,88230.5/
      data rlong.sinlat,coslat,albedo/0.00,0.587,0.810,8.05e-02/
      data vent/178.22/
c
c     Compute declination angle from day and month
c
      call CONRAD (declsc,day,month)
c
c     Compute intermediate sigma levels (sh)
c
      do 9100 k = 1, nk-1
         sh(k)   = sqrt ( s(k)*s(k+1) )
 9100 continue
      sh(nk)      = sqrt ( s(nk) )
c
c     Compute Coriolis parameter
c
      f           = 2.*7.29e-05*sinlat
c
c     Define moisture constants
c
      call WETCNS
c
c     Define the grid point position (not necessary for column model)
c
      ic          = 6
      jc          = 7
c
c     Call radiation routine
```

```
c
      call RAD (rtrsol,rtrter,flxvis,flxiri,ts,t,q,ps,sh,nk,
     &            rlong,sinlat,coslat,s,hz,albedo,vent,declsc,
     &              swr,swc,irtop,ic,jc)
c
c     Write output of shortwave and longwave heating at surface
c     (Watt/m2) and vertical profiles of the shortwave and longwave
c     fluxes(deg/day)
c
      fact       = 86400.
1000  format(2x,2f12.4)
      print *,flxvis,flxiri
      do 9102 k = 1, nk
      write(20,1000)rtrsol(k)*fact,rtrter(k)*fact
9102  continue
      stop
      end
```

10

The Barotropic Model

This chapter basically describes the use of the different numerical techniques presented in this workbook to construct a simple numerical weather prediction model. The illustration here documents a barotropic nondivergent model. An example on the predictive capability of the nondivergent barotropic model over tropical regions is presented in Figure (10.1) (Krishnamurti, 1986).

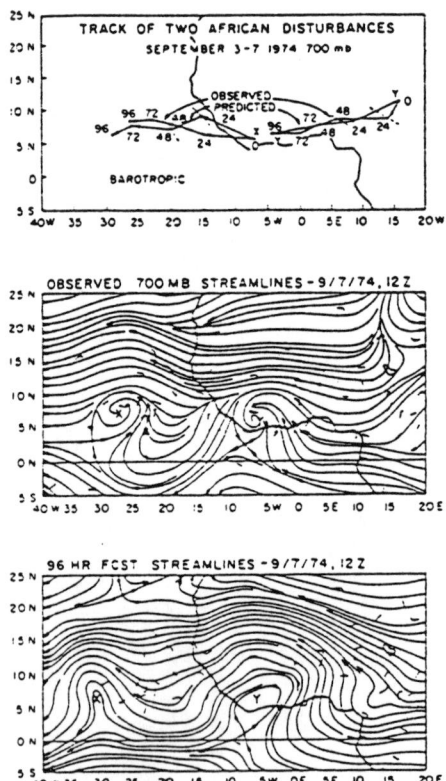

Figure 10.1: Observed and predicted tracks of two African waves for a 5-day period during GATE (1974).

The barotropic model is based on the vorticity conservation concept. Several experiments have been carried out with barotropic models and have shown their usefulness for tropical wind prediction especially at those levels where the divergence is small. Comprehensive documentations on the capability of this model can be found in Krishnamurti et al. (1986).

1. Dynamic of the Barotropic Model

The governing equations for the barotropic model are based on the motion of an inviscid, homogeneous and incompressible fluid on a sphere rotating at the earth's speed. On an (x, y, p) coordinate system, the horizontal motion of the barotropic flow can be written as

$$\frac{\partial u}{\partial t} + u \frac{\partial u}{\partial x} + v \frac{\partial u}{\partial y} = - g \frac{\partial z}{\partial x} + fv \tag{10.1}$$

and

$$\frac{\partial v}{\partial t} + u \frac{\partial v}{\partial x} + v \frac{\partial v}{\partial y} = - g \frac{\partial z}{\partial y} - fu \tag{10.2}$$

where u and v are the two components of the horizontal wind and z is the height of the considered pressure surface. For a nondivergent motion the continuity equation is expressed as

$$\frac{\partial u}{\partial x} + \frac{\partial v}{\partial y} = 0 \tag{10.3}$$

It should be noted that this model is only a crude assumption of the actual atmospheric motion. Nevertheless in the tropics, outside the active weather belt and at the level of nondivergence, the atmospheric flow can be considered to be basically barotropic. Generally, this level is chosen around 700 mb.

The vorticity equation for the barotropic flow may be obtained by differentiating (10.1) with respect to y and (10.2) with respect to x and taking the difference. This leads to an important equation which constitutes an alternative to the set of the above three equations,

$$\frac{\partial \xi}{\partial t} + u \frac{\partial \xi}{\partial x} + v \frac{\partial \xi}{\partial y} + v \frac{\partial f}{\partial y} = 0 \tag{10.4}$$

where $\xi = \frac{\partial v}{\partial x} - \frac{\partial u}{\partial y}$ is the relative vorticity and f represents the

Coriolis parameter. Since f depends only on the latitude, (10.4) may be reformulated to give

$$\frac{d}{dt} \xi_a = 0 \tag{10.5}$$

where $\xi_a = \xi + f$ is the absolute vorticity. This equation simply states that the absolute vorticity is conserved following the parcel's motion. Relating the wind components to the streamfunction as defined in Chapter 4 yields

$$u_\psi = -\frac{\partial \psi}{\partial y} \tag{10.6}$$

and

$$v_\psi = \frac{\partial \psi}{\partial x} \tag{10.7}$$

Equation (10.4) may be expressed as

$$\frac{\partial}{\partial t} \nabla^2 \psi = \frac{\partial \psi}{\partial y} \frac{\partial \nabla^2 \psi}{\partial x} - \frac{\partial \psi}{\partial x} \frac{\partial \nabla^2 \psi}{\partial y} - \beta \frac{\partial \psi}{\partial x} \tag{10.8}$$

or simply

$$\frac{\partial}{\partial t} \nabla^2 \psi = -J(\psi, \nabla^2 \psi) - \beta \frac{\partial \psi}{\partial x} \tag{10.9}$$

where J is the Jacobian operator as defined in Chapter 2 and $\beta = \frac{\partial f}{\partial y}$.

Equation (10.9) forms the basic framework for the barotropic non-divergent model, but it does not give a complete description of the state of the fluid since the pressure field is still not defined. The use of the nonlinear balance equation as defined in Chapter 4 is of great usefulness to close the system,

$$\nabla^2 gz = \nabla \cdot f \nabla \psi + 2J\left(\frac{\partial \psi}{\partial x}, \frac{\partial \psi}{\partial y}\right) \tag{10.10}$$

As described previously, this equation defines the height field from the streamfunction and together with (10.9) constitute a closed system for the evolution of the barotropic nondivergent flow. Equation (10.10) represents actually a nonlinear reverse balance law since it determines the geopotential height from the wind field.

2. Properties of the Barotropic Flow

2.1 Parcel Invariants

It is relatively easy to show that during the evolution of a barotropic nondivergent flow, all powers of the absolute vorticity are conserved following the fluid motion. Multiplying (10.5) by $\frac{1}{n} \xi_a^{n-1}$ one obtains

$$\frac{1}{n} \xi_a^{n-1} \frac{d}{dt} \xi_a = 0 \tag{10.11}$$

or

$$\frac{d}{dt} \xi_a^n = 0 \tag{10.12}$$

In a closed domain one can then track the maximum and minimum values of the absolute vorticity during the time integration of the nondivergent barotropic model.

2.2 Domain Invariants

The nondivergent barotropic flow conserves a number of quantities, the most important of which are known as the quadratic invariants. That is, the mean squared absolute vorticity and the mean kinetic energy per unit mass. It should be noted, however, that the domain average of all powers of the absolute vorticity remain invariant in this model. To show this, (10.12) is rewritten in flux form as

$$\frac{\partial}{\partial t} \xi_a^n = -\frac{\partial}{\partial x} (u \, \xi_a^n) - \frac{\partial}{\partial y} (v \, \xi_a^n) \tag{10.13}$$

Upon integration over a closed domain, one obtains

$$\iint_D \frac{\partial}{\partial t} \xi_a^n \, dxdy = \iint_D \nabla \cdot (\vec{v} \, \xi_a^n) \, dxdy = 0 \tag{10.14}$$

thus,

$$\frac{\partial}{\partial t} \iint_D \xi_a^n \, dxdy = 0 \tag{10.15}$$

The kinetic energy can be obtained simply by multiplying (10.1) by u and (10.2) by v and adding the resulting equations. This leads to

$$\frac{\partial k}{\partial t} = - \vec{v}.\nabla \ (k + gz) \tag{10.16}$$

where $k = (u^2+v^2)/2$ denotes the kinetic energy. Writing this equation in a flux form and integrating it over a closed domain yields

$$\frac{\partial}{\partial t} \iint_D k \ dxdy = 0 \tag{10.17}$$

The domain averaged squared vorticity and kinetic energy are generally known as the barotropic model quadratic invariants. An important conclusion resulting from these conservations is that the mean wavenumber is also conserved. This places a constraint on the spectral distribution of the kinetic energy and does not allow for systematic one way cascade of kinetic energy towards either edge of the spectrum.

3. Barotropic Energy Exchange

An important mechanism for the generation and maintenance of disturbances in barotropic dynamic consists of conversion of the shear vorticity into curvature vorticity. The absolute vorticity conservation implies that during the motion there is an exchange between shear vorticity, curvature vorticity and earth vorticity. The barotropic instability concept is a simple linearization of these exchanges. The principal energy exchange in barotropic dynamics is the conversion of zonal-kinetic energy into eddy kinetic energy and vice versa. Using (10.1) and (10.2), these energy exchanges may be expressed as

$$\frac{\overline{\partial K_e}}{\partial t} = < k_z \cdot k_e > = [u] \ \frac{\partial}{\partial y} \ \overline{[u'v']} \tag{10.18}$$

$$\frac{\overline{\partial K_z}}{\partial t} = < k_e \cdot k_z > = [u] \ \frac{\partial}{\partial y} \ \overline{[u'v']} \tag{10.19}$$

and

$$\frac{\partial \overline{K_t}}{\partial t} = \frac{\overline{\partial K_e}}{\partial t} + \frac{\overline{\partial K_z}}{\partial t} \tag{10.20}$$

where the symbols [ ] and —— denote the zonal and meridional averages, respectively and the prime represents a deviation from the mean (Krishnamurti, 1979). It is useful to monitor these energy exchanges during the integration of the barotropic model and to examine the evolution of the eddies.

4. Model Structure and Boundary Conditions

As mentioned earlier the barotropic model is based on the integration of Equations (10.9) and (10.10) with appropriate boundary conditions. The numerical technique consists of writing the governing equations in finite differences analogs at a discrete set of points in space and time. The equations are solved on a regularly spaced horizontal grid.

One of the problems inherent to limited area grid point models is the specification of the boundary conditions. The solution of the governing equations depends continuously on the boundary values which makes the problem very sensitive to their specification. The most important problem resulting from the boundaries treatment is the contamination of the forecast by inertial gravity waves reflected into the domain from the boundaries. For the barotropic nondivergent model, the boundary conditions specification does not pose a major problem since inertial gravity waves are not generated. In this case a cyclic continuity in the zonal direction is imposed on the initial field and is maintained throughout the integration. Basically this consists of extending the domain of integration by adding some extra grid points in the zonal direction. At the north-south boundaries, different specifications may be applied. For example one can use

$$\psi(x, y_B, t) = \psi(x, y_B, t_0) \tag{10.21}$$

or

$$\psi(x, y_B, t) = \int_D \psi(x, y_B, t_0) / \int_D dx \tag{10.22}$$

where t_0 is the initial time. In this model (10.21) is used along the northern and southern boundaries. This specification basically leaves the north-south boundaries open. The boundary conditions used in the solution of Poisson's type of equations are similar to those used for the relaxation method in Chapter 4.

5. Treatment of the Advective Terms and Time Differencing Scheme

In order to avoid the generation of unbounded vorticity and kinetic energy during the model integration, it is imperative that the integral properties of the barotropic nondivergent model be conserved in the finite difference analogues of the predictive equations. Arakawa's Jacobian developed in Chapter 2 satisfies these constraints and is utilized in the treatment of the advective terms of the model.

The time differencing scheme used in this model is the Matsuno backward described in Chapter 2. This scheme is computationally stable and produces a little damping. However, because of its two step iterations, the scheme takes about twice as long as the leap-frog. Generally, Matsuno's scheme is not used for the integration of an operational model, but is used periodically during model integrations to suppress computational modes arising from the leap-frog scheme.

6. Initial Conditions

The finite difference form of the predictive equation (10.9) requires a specification of an initial streamfunction field to start the forecast. This is simply done through the solution of the following Poisson equation,

$$\xi = \nabla^2 \psi \tag{10.23}$$

Given the wind components, the relative vorticity is computed using the centered difference finite approximation,

$$\xi_{ij} = \frac{v(i+1,j) - v(i-1,j)}{2\Delta x} - \frac{u(i,j+1) - u(i,j-1)}{2\Delta x} \tag{10.24}$$

Cyclic continuity is used in the east-west direction. Values at the northern and southern boundaries are obtained using linear extrapolation. The initial streamfunction field is then obtained as a solution of (10.23) with boundary conditions similar to those used in Chapter 4. The vorticity is obtained from the wind field using (10.24).

7. Description of the Code

The code for the nondivergent barotropic model provided in this chapter consists essentially of three programs, each of which performs a specific job. The model is designed with simplicity and flexibility in mind.

The three parts of the model can be run separately but only in one logical sequence since output from one program is used for subsequent program. The first program called (*INFIELD*) is designed to prepare the initial streamfunction from the wind field. This part also defines the boundary conditions and some other constants needed by the model. This program can be used either to prepare the initial field for the barotropic nondivergent model or for the shallow water model described in Chapter 11. In the latter case, after the computation of the streamfunction is completed, the program evaluates the height field using the nonlinear reverse balance law. Furthermore, the program provides the option of initializing the shallow water model with or without topography. Therefore, caution is recommended when selecting the different initialization options which are

IOPT1 = 0	IOPT2 = 0	initialize barotropic model
IOPT1 = 1	IOPT2 = 0	initialize shallow water model without terrain
IOPT1 = 1	IOPT2 = 1	initialize shallow water model with terrain

In case the terrain field is used in the shallow water model, it needs to be provided as an input.

The second program called (*BARO*) performs the integration of the barotropic model. Before starting the integration, the model calls subprogram *INIT* to define some control parameters such as the total forecast length, the time step of integration and the output time interval. It also defines the limits of the integration domain as well as some other arguments required to run the model. The model monitors the barotropic energy exchanges through subroutine *ENERGY* and the extreme values of the potential vorticity.

The third part of the model (*BAROUT*) reads the forecast's streamfunction field and computes the wind components. The geopotential height is then deduced using the nonlinear balance relation with a geostrophic balance assumed at the meridional boundaries.

Each part of these three codes is provided and is internally well documented. Sequences of the entire execution are described in Figure (10.2).

An example of implementation of the barotropic model with the code provided is presented in this section. Figure (10.3) shows the initial streamlines field over the considered domain. The wind field obtained after 12 hour integration is shown in Figure (10.4).

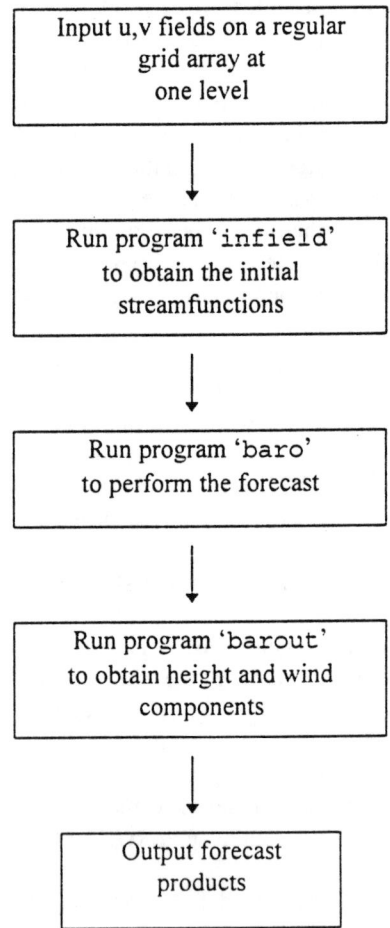

Figure 10.2: Sequences of execution of the barotropic model.

program *INFIELD*

```
c
c      This program can be used either to prepare the
c      initial streamfunction field for the nondivergent
c      barotropic model forecast or the u,v,z fields for
c      the shallow water model iterative initialization/forecast.
c      If the program is to be used for preparing the initial
c      field for the barotropic model, set parameters
c          iopt1  =  0, and
c          iopt2  =  0
c      In this case the program reads in the initial wind
c      fields from unit 21 and outputs the streamfunction
c      field to unit 22.
c      If the program is to be used for preparing the initial
c      field for the shallow water model without terrain,set
c          iopt1  =  1, and
c          iopt2  =  0
c      The program will read in the initial u, v fields from
c      unit 21 and outputs the u,v,z fields to unit 22.
c      For the preparation of initial field  for shallow water
c      model with terrain, set
c          iopt1  = 1, and
c          iopt2  = 1 .
c
c      input      : (1)    Initial grid point zonal and meridional
c                          components of the wind fields from unit 21.
c                 (2)    if program is used for preparing the initial
c                          fields for the shallow water model with
c                          terrain, then the reduced terrain field and
c                          its zonal and meridional gradients are read
c                          in from unit 20.
c
c      output     : (1)    Logical unit 22 contains the output fields
c                          described above.
c                 (2)    logical unit 10 is the dayfile file.
c                 (3)    logical unit 23 contains the terrain field
c                          and its zonal meridional gradients.
c
c      Definition :
c
c      l           : East-west dimension of the domain
c      m           : North-south dimension of the domain
c      ladd        : Number of points added to make the domain cyclic
```

```
c    u,v         : Horizontal wind components
c    psi         : Streamfunction
c    z           : Height field
c    cor         : Coriolis parameter for each latitude
c    dx          : East-West grid spacing for each latitude
c    dx2         : East-West grid spacing for terrain field
c    aaa,uu,vv   : Work arrays
c    wrka1,wrka  : Work arrays
c    hd,dhdy,dhdx: Terrain and its gradients
c
c
     parameter(l= 45,m = 21,ladd = 6,lcy=l+ladd,iopt1 = 0,iopt2 = 0 )
     common  u(l,m)      , v(l,m)       , psi(l,m)
     real     aaa(lcy,m) , uu(lcy,m)    , vv(lcy,m)
     real     z(lcy,m)   , wrka1(lcy,m), wrka(l,m)
     real     cor(m)     , dx(m)        , dx2(m)
     real     dhdy(lcy,m), dhdx(lcy,m) , hd(lcy,m)
     equivalence (wrka,vv)
c
c    Open the input-output files. Unit 20 and 23 are opened only
c    if the terrain is used.
c
     open (21,file='uv.dat',status='old')
     open (22,file='infield_out.dat',status='unknown')
     if    (iopt2.ne.0) then
     open (20,file='terrain.dat',status='old')
     open (23,file='terrain.out',status='unknown')
     endif
c
c    Print model functions
c
     print *,' '
     print *,'    model status'
     print *,' '
     if (iopt1.eq.0.and.iopt2.eq.0) then
     print *,'    initialization for barotropic model '
     print *,' '
     print *,'    the model reads the initial u,v field from unit 21'
     print *,'    and outputs the streamfunction field to unit 22'
     print *,' '
     else
     if (iopt1.eq.1.and.iopt2.eq.0) then
     print *,'    initialization for the single level '
```

```
                  print *,'    primtive equation model without terrain'
                  print *,' '
                  print *,'    the model reads in the input u,v fields from unit 21'
                  print *,'    and outputs the u,v,z fields to unit 22'
                  print *,' '
                  else
                  if (iopt1.eq.1.and.iopt2.eq.1) then
                  print *,'    initialization for the single level '
                  print *,'    primtive equation model with terrain'
                  print *,' '
                  print *,'    the model reads in the input u,v fields from unit 21'
                  print *,'    and outputs the u,v,z fields to unit 22'
                  print *,'    the initial terrain field is read from unit 20'
                  print *,'    and the output field is written to unit 23'
                  print *,' '
                  endif
                  endif
                  endif
                  print *,' '
                  print *,'                    NOTE :
                  print *,'    the grid size and/or the southernmost latitude need'
                  print *,'    to be changed in subroutine "const" if the
                  print *,'    domain/or data is changed'
                  print *,' '
                  print *,' '
      c
      c    Rewind all input tapes.
      c
                  rewind (20)
                  rewind (21)
                  rewind (22)
      c
                  l1          = l-1
                  m1          = m-1
                  m2          = m-2
      c
      c    Call subroutine CONST to define the constants
      c    required by the program.
      c
                  call CONST (m,dx,dy,cor,beta,slat,dphi)
      c
      c    Read in the initial u, v fields.
      c
```

```
            read (21,878) ((u(ii,jj),ii=1,l),jj=1,m)
            read (21,878) ((v(ii,jj),ii=1,l),jj=1,m)
  878       format(5e13.6)
c
c     Print first and last elements of u and v arrays as a check.
c
            write (6,1000) u(1,1),u(l,m)
            write (6,1001) v(1,1),v(l,m)
            write (6,1002)
 1000       format(1x,'u(1,1)=',e12.4,5x,'u(l,m)=',e12.4)
 1001       format(1x,'v(1,1)=',e12.4,5x,'v(l,m)=',e12.4)
 1002       format(//)
 1003       format(1x,'z(1,1)=',e12.4,5x,'z(l,m)=',e12.4)
c
c     Call subroutine STREAMF to compute the stream-
c     function field.
c
            call STREAMF (u,v,l,lcy,l,m,dy,dx,wrka,psi)
c
c     Call subroutine CYCLE to puts east-west cyclic
c     boundary condition on the streamfunction field.
c
            do 10100 j = 1, m
            do 10100 i = 1, l
               aaa(i,j)= psi(i,j)
10100       continue
            call CYCLE (aaa,lcy,m,1)
            if (iopt1.eq.0) then
            write (22,879) ((aaa(i,j),i=1,lcy),j=1,m)
  879       format(6e13.6)
            else
c
c     Call subroutine ZFIELD to compute the height field
c     from the streamfunction field.
c     the mean height field of the 700 mb surface is considered.
c
            zbar      = 3000.
            zbar2     = 0.
            call ZFIELDMOD(aaa,uu,vv,lcy,m,dx,dy,cor,
     &                          wrka1,z,1.,1.,zbar,beta)
            do 10102 j= 1, m
            do 10102 i= 1, lcy-1
               zbar2  = z(i,j) + zbar2
```

```
10102   continue
        zbar2     = zbar2/(m*(lcy-1))
        do 10104 j= 1, m
        do 10104 i= 1, lcy
           z(i,j)   = z(i,j)-zbar2+zbar
10104   continue
c
c       Call subroutine CYCLE to put east-west cyclic conditions on
c       the u,v fields.
c
        do 10106 j= 1, m
        do 10106 i= 1, 1
           aaa(i,j)= u(i,j)
10106   continue
        call CYCLE (aaa,lcy,m,1)
        write (22,879) ((aaa(i,j),i=1,lcy),j=1,m)
        do 10108 j= 1, m
        do 10108 i= 1, 1
        aaa(i,j)    = v(i,j)
10108   continue
        call CYCLE (aaa,lcy,m,1)
        write (22,879) ((aaa(i,j),i=1,lcy),j=1,m)
c
c       Read the terrain field if it is used. Read in terrain
c       field at 1 by m and then run CYCLE to get cyclic values
c       between 1 and lcy in zonal direction
c
        if (iopt2.eq.1) then
        read (20,880) ((wrka1(ii,jj),ii=1,1),jj=1,m)
        call CYCLE (wrka1,lcy,m,1)
        aamax     = wrka1(1,1)
c
c       Compute the maximum value of the terrain field and
c       normalized it.
c
        do 10110 j= 1, m
        do 10110 i= 1, lcy
10110   if (wrka1(i,j).gt.aamax) aamax = wrka1(i,j)
        do 10112 j= 1, m
        do 10112 i= 1, lcy
10112   wrka1(i,j) = (wrka1(i,j)/aamax) * 1000.0
c
c       Subroutine TERR create tape21 for shallow water
```

```
c     model when iopt2.ne.0
c
      call TERR (wrka1,lcy,m,dphi,slat,dhdy,dhdx,dx2,hd)
 880  format(6e13.6)
      do 10114 j= 1, m
      do 10114 i= 1, lcy
10114    z(i,j)  = z(i,j)-1.0*wrka1(i,j)
      endif
c
c     Write output fields
c
      write (22,879) ((z(i,j),i=1,lcy),j=1,m)
      write (6,1000) u(1,1),u(l,m)
      write (6,1001) v(1,1),v(l,m)
      write (6,1003) z(1,1),z(l,m)
      write (6,1002)
      endif
      stop
      end




      program BARO
c
c     This program performs the nondivergent barotropic
c     model forecast. The prediction equation uses the
c     the streamfunction as predictive variable.
c     The Eulerian frame is used. The advective term
c     is  computed through the 9 point Arakawa jacobian.
c     Time integration is performed using the Matsuno
c     time scheme. Some properties of the nondivergent
c     barotropic fluid, for example the maximum absolute
c     vorticity and the domain mean kinetic energy,
c     are also computed.
c
c     Input   :
c                    streamfunction on unit 23. This file is
c                    read in subroutine INIT.
c
c     Output  :
c                    (i)  forecast streamfunctions at preset
c                    intervals of nint hours (see comments on
c                    symbols below). These are written onto unit
```

```
c                17. The first record is the
c                streamfunction field at hour 0. Subsequent
c                records contain the forecast records
c                which are sequentially written according
c                to the forecast hour.
c
c                (ii) comment statements on energy computations
c                and vorticity parameters during time integration.
c                These are written onto unit 6 and normally
c                this file is assigned to the screen.
c
c    Definition :
c
c    l          : number of grid points in the zonal
c                   direction.
c    m          : number of grid  points in the meridional
c                   direction.
c    n          : number of vertical levels  (1 in this
c                   case).
c    l1         : l-1
c    m1         : m-1
c    n1         : n-1
c    l2         : l-2
c    m2         : m-2
c    ladd       : number of grid points added to make
c                   domain cyclic.
c    phi(m)     : latitudes of each row of grid points
c    cor(m)     : Coriolis parameter for each row of grid
c                   points.
c    dy         : grid size in the meridional direction
c                   in meters.
c    dx(m)      : grid  size  in  the  zonal  direction in
c                   meters for each row of grid points.
c    dt         : time step  of integration in seconds.
c    nint       : time interval in hours to output
c                   forecast products. This is later
c                   modified to represent the number
c                   of time steps before the next
c                   forecast products are output to file
c                   17.
c    tend       : Total length in hours of the forecast.
c    asml(n)    : smallest value of the absolute vorticity
c    alrg(n)    : largest value of the absolute vorticity.
```

```
c     psi(l,m,n)   : streamfunction field.
c     a1(l,m,n)    : array containing either the relative
c                     or absolute vorticity.
c     a2(l,m,n)    : another working array containing either
c                     the relative or absolute vorticity.
c     b1(l,m,n)    : array containing the advective term,
c                     jacobian of (psi,absolute vorticity).
c
c     Subroutines
c     called       : bound,energy,equal,init,jac,mod,
c                     lap,large,relaxt,small,vort
c
      parameter( l = 51,m = 21,n = 1)
c
c
      common     l1,        m1,        l2,        m2,
     &           ladd,      phi(m),    cor(m),    dy,
     &           dx(m),     dt,        nint,      mint,
     &           tend,      time,      asml(n),   alrg(n),
     &           psi(l,m,n), a1(l,m,n), a2(l,m,n), b1(l,m,n),
     &           slat       dphi
c
      write(6,1000)
c
c     Call subroutine INIT to read in initial parameters
c     and to define the initial state.
c
      call INIT
c
c     Write the initial state to output file.
c
      do 10200 k = 1, n
      write (17,878) ((psi(i,j,k),i=1,l),j=1,m)
10200 continue
 878  format(6e13.6)
c
c     Counts the number of timesteps in nrst.
c
      nrst       = 0
c
c     The largest and smallest value of the absolute vorticity in
c     each time step are calculated by subroutines LARGE and
c     SMALL, respectively.
```

```
c
          call LARGE (a1,l,m,n,alrg)
          call SMALL (a1,l,m,n,asml)
          write(6,1001)nrst+1
          write(6,1002) time,alrg(1),asml(1)
 1001    format(10x,'number of time steps = ',i3)
 1002    format(/,5x,'time =',f8.0,4x,'max,min abs vort =',e11.4,
      &              5x,e11.4,/)
c
c     The energy parameters between 2 latitudes at time = 0
c     is calculated using subroutine ENERGY.
c
          jslat1        = 2
          jnlat1        = m1
          jslat2        = 6
          jnlat2        = m-jslat2+1
c
          call ENERGY (psi,l,m,n,l1,dy,dx,time,jslat1,jnlat1,slat,dphi)
          call ENERGY (psi,l,m,n,l1,dy,dx,time,jslat2,jnlat2,slat,dphi)
c
c     Time integration following the Matsuno time scheme
c     is carried out. The scheme consists of two steps,
c     namely the  predictor step and the corrector step.
c
c     Predictor Step :
c
c     First the advective term  i.e. the jacobian
c     of psi and delsquared of (psi + f ) is computed.
c
   13     call JACMOD (b1,psi,a1,dx,dy,l,m,n,l1,m1,l2,m2,ladd)
c
c     The time integration for the predictor step is
c     carried out next.  This gives the first guess relative
c     vorticity field. It is stored in array a2.
c
          do 10202 j = 1, m
          do 10202 i = 1, l
10202    a2(i,j,1)     = (a1(i,j,1)-b1(i,j,1)*dt-cor(j))
          do 10204 i = 1, l
          a2(i,1,1)     = 0.
10204    a2(i,m,1)     = 0.
c
c     Corrector Step  :
```

```
c
c      Relaxation of a2 is done to obtain the first guess psi field.
c      With this, the first guess absolute vorticity  field (a2+f),
c      and the first guess advective term, J(psi,delsquared (psi + f))
c      are computed.
c
       call RELAXT (psi,a2,dx,dy,l,l,m,l1,m1,m2,1)
       call VORT    (a2,cor,l,m,n)
       call JACMOD     (b1,psi,a2,dx,dy,l,m,n,l1,m1,l2,m2,ladd)
c
c      The time integration for the corrector step is carried out
c      here. The second guess relative vorticity field is stored
c      in array a2.
c
       do 10206 j = 1, m
       do 10206 i = 1, l
10206  a2(i,j,1)     = (a1(i,j,1)-b1(i,j,1)*dt-cor(j))
       do 10208 i = 1, l
       a2(i,1,1)     = 0.
10208  a2(i,m,1)    = 0.
c
c      a2 is relaxed to obtain the second guess psi field. The second
c      guess absolute vorticity field (a2+f), is calculated and
c      transfered into array a1 by subroutine equal for the next
c      time step computations.
c
       call RELAXT (psi,a2,dx,dy,l,l,m,l1,m1,m2,1)
       call VORT    (a2,cor,l,m,n)
       call EQUAL   (a1,a2,l,m,n)
c
c      The time integration counter, time, is increased by
c      timestep dt.
c
       nrst      = nrst + 1
       time      = time + dt
       write(6,1001)nrst+1
c
c      The largest and smallest value of the absolute vorticity in
c      each time step are calculated by subroutines LARGE and
c      SMALL, respectively.
c
       call LARGE (a2,l,m,n,alrg)
       call SMALL (a2,l,m,n,asml)
```

```
             write(6,1002) time,alrg(1),asml(1)
c
c        The energy parameters between 2 latitudes for each time step
c        are calculated using subroutine ENERGY
c
             call ENERGY (psi,l,m,n,11,dy,dx,time,jslat1,jnlat1,slat.dphi)
             call ENERGY (psi,l,m,n,11,dy,dx,time,jslat2,jnlat2,slat.dphi)
c
c        The process  leading to another time integration
c        is repeated. However, if time = nint (the preset
c        output interval), the psi field is written into
c        the output file. If time = tend (the preset time
c        to end forecast), the program execution is terminated.
c
             if (nrst-nint) 13,14,13
     14      continue
             do 10210 k = 1, n
  10210  write (17,878) ((psi(i,j,k),i=1,l),j=1,m)
             nint      = nint+mint
c
             print *,'time ,tend',time,tend
             if (time-tend) 13,16,16
     16      continue
   9998      continue
   1000      format(/)
             stop
             end




             program BAROUT
c
c        This program reads in the initial and forecast
c        streamfunction fields and computes the components
c        of the wind field. The option of computing the height
c        field is also built in. The height field is deduced
c        through the nonlinear reverse balance law.
c        Geostrophic balance is assumed for the southern and
c        northern boundaries. The height field is not written
c        out to the output field. Wind and height fields are
c        computed within subroutine ZFIELD.
c
c        input           :
```

```
c
c      (1) initial  and forecast streamfunction fields from
c      logical unit 17.
c
c      output       :
c
c      (1) logical unit 23 contains the initial and
c      sequences of forecast u,v fields for each forecast hour.
c
c      subroutines  : bound, const, jac, lap, relax, zfield.
c      called
c
c      parameter nmap defines the number of  sets of
c      streamfunction fields to be processed.
c
c      tend   : is the forecast time of BARO
c      nint   : is the output times from BARO
c      nmap : is derived from the above 2 variables
c
           parameter(l = 51, m = 21)

           real  psi(l,m),u(l,m),v(l,m)
           real  z(l,m),wrka(l,m),cor(m),dx(m)

           m1            = m-1
           m2            = m-2
           tend          = 24
           itend         = int(tend)
           nint          = 12
           nmap          = tend/nint + 1
           print *,' number of output psi ',nmap
           call CONST  (m,dx,dy,cor,beta,slat,dphi)
c
           open(17,file='baro_out.dat',status='old')
           open(23,file='ufield_out',status='unknown')
c
           do 10300 n = 1, nmap
c
c      Read the streamfunction field.
c
           read(17,878) ((psi(i,j),i=1,l),j=1,m)
  878      format(6e13.6)
c
```

```
c     Subroutine 'zfield' is called to compute the
c     height field from the streamfunction using
c     non-linear balance. The arrays u(i,j) and v(i,j)
c     contain the components u and v of the wind field.
c
        zbar        = 3000.
        zbar2       = 0.
        call ZFIELD (psi,u,v,l,m,dx,dy,cor,wrka,z,1.,1.,zbar,beta)
        do 10302 j = 1, m
        do 10302 i = 1, l-1
          zbar2     = zbar2 + z(i,j)
10302   continue
        zbar2       = zbar2 / (m*(l-1))
        do 10304 j = 1, m
        do 10304 i = 1, l
          z(i,j)    = z(i,j)-zbar2 + zbar
10304   continue
c
c     The southern and northern boundary values of the
c     u and v fields are set by linear interpolation.
c
        do 10306 i = 1, l
          u(i,1)  = 2.*u(i,2)  - u(i,3)
          v(i,1)  = 2.*v(i,2)  - v(i,3)
          u(i,m)  = 2.*u(i,m1) - u(i,m2)
          v(i,m)  = 2.*v(i,m1) - v(i,m2)
10306   continue
c
c     Write u,v components to unit 23.
c
        write (23,1000) ((u(i,j),i=1,l-6),j=1,m)
        write (23,1000) ((v(i,j),i=1,l-6),j=1,m)
10300   continue
 1000   format(6e13.6)
        stop
        end
```

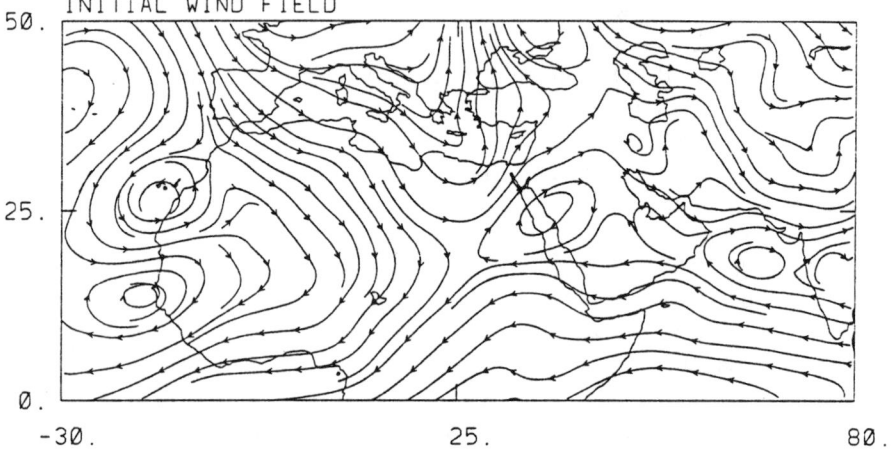

Figure 10.3: Initial wind field at 700 mb for the initialization of the barotropic model (May 14, 1985 1200 GMT).

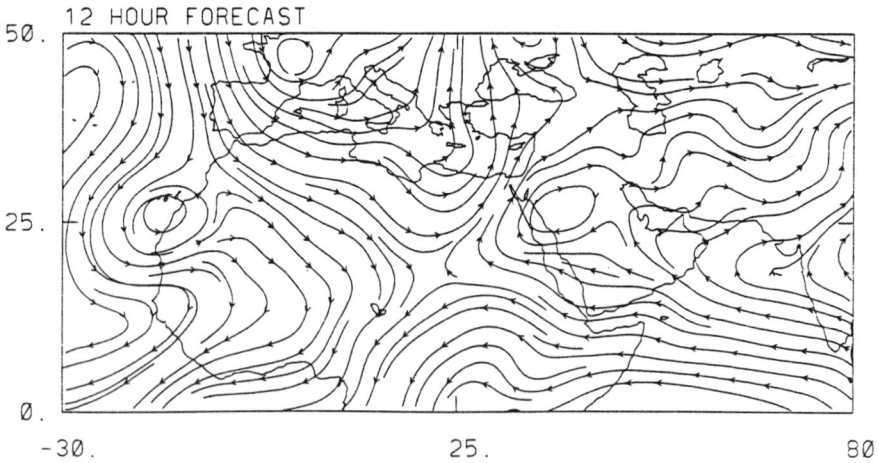

Figure 10.4: 12 hour barotropic forecast at 700 mb starting from May 14, 1985 at 1200 GMT.

11

The Single Level Primitive Equations Model

This model is generally considered better than the barotropic model in the sense that it allows for divergence to occur at its free surface. It is a single level model and is based on the primitive equations. The model is also known as the Laplace's tidal equations or the barotropic divergent model, but it is most commonly referred to as the shallow water equations model.

1. Dynamic of the Single Level Primitive Equations Model

The shallow water model is governed by the horizontal equations of motion and the mass continuity equation. These are expressed as

$$\frac{\partial u}{\partial t} + u \frac{\partial u}{\partial x} + v \frac{\partial u}{\partial y} = -g \frac{\partial(z+h)}{\partial x} + fv \tag{11.1}$$

$$\frac{\partial v}{\partial t} + u \frac{\partial v}{\partial x} + v \frac{\partial v}{\partial y} = -g \frac{\partial(z+h)}{\partial x} - fu \tag{11.2}$$

$$\frac{\partial z}{\partial t} + u \frac{\partial z}{\partial x} + v \frac{\partial z}{\partial y} = -z \left[\frac{\partial u}{\partial x} + \frac{\partial v}{\partial y} \right] \tag{11.3}$$

where u and v are the wind horizontal components, h represents the topographic height and z is the depth of the free surface measured from the top of the topography (Figure 11.1). Since the model allows for divergence, one of its solutions would be inertial-gravity waves. These fast moving waves can amplify rapidly during the model integration and contaminate the forecast. A special treatment is then performed to balance the initial wind and pressure fields.

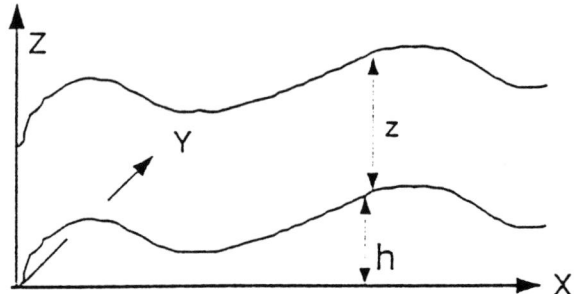

Figure 11.1: Schematic diagram of the shallow water model.

2. Properties of the Single Level Primitive Equations Model

2.1 Parcel Invariants

Unlike the barotropic nondivergent model which conserves the relative vorticity, the shallow water model is based on the potential vorticity conservation. This can be shown by forming the vorticity equation from (11.1) and (11.2) as

$$\frac{\partial}{\partial t}(\xi + f) = -\ \vec{V}.\nabla(\xi + f) - (\xi + f)\ \nabla.\vec{V} \tag{11.4}$$

Using the continuity equation and eliminating the divergence term leads to

$$\frac{D}{Dt}\ \xi_p = 0 \tag{11.5}$$

where $\xi_p = \dfrac{\xi + f}{z}$ is the potential vorticity and the operator $\dfrac{D}{Dt}$ is defined as

$$\frac{D}{Dt} = \frac{\partial}{\partial t} + \vec{V}.\nabla \tag{11.6}$$

Similarly, multiplying (11.5) by $\dfrac{1}{n}\ \xi_p^{n-1}$ leads to

$$\frac{D}{Dt} \xi_p^n = 0 \tag{11.7}$$

which shows that all powers of the potential vorticity are conserved following the parcel motion during the integration of the shallow water equations.

2.2 Domain Invariants

Among other domain invariants, the shallow water model conserves the mean free surface depth z and $\ln(z^2/2)$. An important invariant of this system is, however, the total energy parameter defined as

$$E_p = z (k + g(\frac{z}{2} + h)) \tag{11.8}$$

This can be shown by writing the kinetic energy evolution from the momentum equations,

$$\frac{\partial k}{\partial t} + \overrightarrow{V} . \nabla k = - \overrightarrow{V} . \nabla g(z + h) \tag{11.9}$$

Upon multiplication of (11.3) by (k + gz + gh) and (11.9) by z and adding the resulting equations, one obtains

$$\frac{\partial}{\partial t} z(k + \frac{gz}{2} + gh) + \nabla . kz \overrightarrow{V} + \nabla . gz(z + h) \overrightarrow{V} = 0 \tag{11.10}$$

Finally, the integration of this equation over a closed domain leads to

$$\frac{\partial}{\partial t} \oint z(k + \frac{gz}{2} + gh) \, dx \, dy = 0 \tag{11.11}$$

stating that the domain averaged energy parameter is conserved during the integration of the model. These domain invariant quantities constitute important model diagnostics.

3. Model Structure and Boundary Conditions

The single level primitive equations model is integrated over a regular latitude-longitude grid similar to that used for the barotropic nondivergent model. The boundary conditions specification is also similar to that used in the barotropic model.

4. Treatment of the Advective Terms and Time Differencing Scheme

The conservation of the integral properties requires a judicious choice of the advective scheme in any numerical weather prediction model. The single level primitive equations model presented here makes use of the semi-Lagrangian technique as described by Mathur (1970). This is based on the following principle, if A, B, C represent the forcing functions of the zonal and meridional momentum equations and the continuity equation, respectively, the governing equations can be written in a Lagrangian framework as

$$\frac{Du}{Dt} = A \tag{11.12}$$

$$\frac{Dv}{Dt} = B \tag{11.13}$$

$$\frac{Dz}{Dt} = C \tag{11.14}$$

Using the Matsuno backward time differencing scheme, the advective terms are estimated using the following method. First assume that the parcel which is located at point P at time t reaches the point Q at time $t+\Delta t$ (Figure 11.2). Since the position of P is unknown, successive approximations are made using the grid point value of Q at time t. If x and y represent the zonal and meridional distances of P from Q, then using Taylor's expansion, a first guess position for P is obtained,

$$x = - u \big|_{Q_t} \Delta t - \frac{1}{2} A \big|_{Q_t} \Delta t^2 \tag{11.15}$$

$$y = - v \big|_{Q_t} \Delta t - \frac{1}{2} B \big|_{Q_t} \Delta t^2 \tag{11.16}$$

With the first guess position of P, first guess values of u, v, A and B at P are determined using a 9-point Lagrangian interpolation scheme as

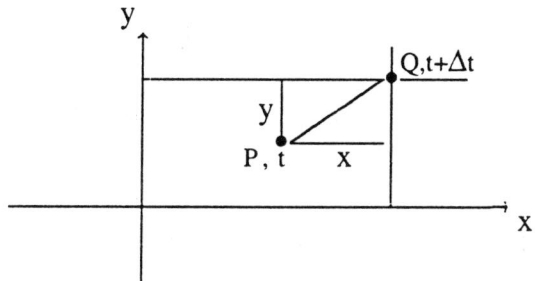

Figure 11.2: Semi-Lagragian scheme.

defined in Chapter 5, Section 2. With the first guess values at time t, the second guess position of P is obtained as follows,

$$x = -u\big|_{P_t} \Delta t - \frac{1}{2} A\big|_{P_t} \Delta t^2 \qquad (11.17)$$

$$y = -v\big|_{P_t} \Delta t - \frac{1}{2} B\big|_{P_t} \Delta t^2 \qquad (11.18)$$

The new position of P allows then the obtention of u, v, z, A, B and C at P and time t. The u, v, z values at grid point Q and time $t + \Delta t$ are computed using the predictor-corrector method of the Matsuno time integration scheme. The predictor step consists of a trial forward step given by

$$U_{Q_{t+\Delta t}} = u\big|_{P_t} + A\big|_{P_t} \Delta t \qquad (11.19)$$

$$V_{Q_{t+\Delta t}} = v\big|_{P_t} + B\big|_{P_t} \Delta t \qquad (11.20)$$

$$Z_{Q_{t+\Delta t}} = z\big|_{P_t} + C\big|_{P_t} \Delta t \qquad (11.21)$$

The correction consists of a backward time differencing step and is expressed as

$$U_{Q_{t+\Delta t}} = u\big|_{P_t} + A\big|_{Q_{t+\Delta t}} \Delta t \qquad (11.22)$$

$$V_{Q_{t+\Delta t}} = v\Big|_{P_t} + B\Big|_{Q_{t+\Delta t}} \Delta t \tag{11.23}$$

$$Z_{Q_{t+\Delta t}} = z\Big|_{P_t} + C\Big|_{Q_{t+\Delta t}} \Delta t \tag{11.24}$$

5. Computation of the Forcing Functions

In the computation of the forcing functions, the pressure gradient force term in the momentum equations and the divergence term in the continuity equation are approximated through the standard centered differences. This method of approximation gives higher order accuracy but leads to splitting of solutions that are associated with 2 grid distance waves. The splitting is due to the use of a nonstaggered horizontal grid structure for the specification of the predicted variables. To overcome this, a correction term is applied to the momentum and continuity equations as

$$\frac{Du^{n+1}}{Dt} = -g\frac{\partial(z+h)^{n+1}}{\partial x} + fv^{n+1} + g\Delta t\left[-z^n\frac{\partial^2 u^n}{\partial x^2}\right]_{in} + g\Delta t\left[z^n\frac{\partial^2 u^n}{\partial x^2}\right]_{co} \tag{11.25}$$

$$\frac{Dv^{n+1}}{Dt} = -g\frac{\partial(z+h)^{n+1}}{\partial y} + fu^{n+1} + g\Delta t\left[-z^n\frac{\partial^2 v^n}{\partial y^2}\right]_{in} + g\Delta t\left[z^n\frac{\partial^2 v^n}{\partial y^2}\right]_{co} \tag{11.26}$$

and

$$\frac{Dz^{n+1}}{Dt} = -z\left[\frac{\partial u}{\partial x} + \frac{\partial v}{\partial y}\right]^{n+1} + g\Delta t\left[\frac{\partial^2 z^n}{\partial x^2} + \frac{\partial^2 z^n}{\partial y^2}\right]_{in} - g\Delta t\left[\frac{\partial^2 z^n}{\partial x^2} + \frac{\partial^2 z^n}{\partial y^2}\right]_{co} \tag{11.27}$$

where the superscript denotes the time step at which the terms are evaluated and subscripts *in* and *co* stand for uncorrected and corrected, respectively. For a variable e^n_{ij}, the uncorrected and corrected terms are evaluated as

$$\left[\frac{\partial^2 e}{\partial x^2}\right]^n_{ij_{in}} = \frac{e^n_{i+2,j} + e^n_{i-2,j} - 2e^n_{ij}}{4\Delta x^2} \tag{11.28}$$

$$\left[\frac{\partial^2 e}{\partial x^2}\right]^n_{i,j_{co}} = \frac{e^n_{i+1,j} + e^n_{i-1,j} - 2e^n_{ij}}{\Delta x^2} \tag{11.29}$$

A detailed description on the derivation of these correction terms may be found in Kanamitsu (1975) and Krishnamurti (1974).

6. Initialization of the Single Level Primitive Equations Model

The difficulty of obtaining a synoptic definition for the pressure-wind relation in low latitude is well known. Furthermore, because of the errors in the vertical temperature profile measurements, the hydrostatic law is not reliable, especially in the lower troposphere where a temperature error of about 1°C can lead to errors on the order of 20 m in the geopotential height. Therefore, the height-wind relation must rely on dynamical methods which are model dependent. The question raised here is how to obtain the geopotential height field in low latitudes regions. In the single level primitive equations model, the height field is deduced from the horizontal motion through different balance laws applied successively in the order of their increasing complexity as described in Chapter 4, Section 3. This procedure is called static initialization. The height field obtained from the nonlinear balance laws is next improved by a process called dynamic initialization. The dynamic initialization procedure is based on the shallow water equations. The process entails a forward-backward integration of the single level primitive equations for u, v and z fields obtained from the static initialization. Sugi (1986) proposed a dynamic normal-mode initialization procedure that utilizes a forward-backward iteration (around the first time step). In this method the forward-backward iterations are carried out with the linear term while keeping the nonlinear forcings fixed. After every 100 linear iterations the nonlinear terms are updated, and this entire process is repeated five times. Sugi (1986) has demonstrated a rapid damping of high-frequency modes with a solution that corresponds closely to that obtained from nonlinear normal mode initialization. Sugi's study was carried out within a global spectral model at a 42 waves triangular truncation. The transform grid at this resolution corresponds approximately to 250 km.

When Sugi's method is carried out in regional grid-point models at a resolution of roughly 50 km or higher, a rapid damping of the gravitational high-frequency oscillations occurs. However, in mesoscale models with still higher resolutions, this iteration scheme converges slowly and requires a reformulation. The present study utilizes grid sizes of the order of 50 km or higher, and the method appears to converge quite rapidly.

In the dynamic normal-mode initialization as formulated by Sugi (1986) the initialization is performed in the physical space; however, it could be easily extended and performed in the vertical-mode space (Kumar, 1990).

The primitive equations are integrated forward and backward with small time steps of the order of few minutes, but satisfying the CFL criteria. The purpose of this is to adjust the motion and pressure fields to an equilibrium state that may depart from the so-called balance laws. In the process of this two way integration, inertial-gravity oscillations are excited and the final state of the iterative initialization would normally vary slowly on forward integration. The adjusted wind field varies slightly from the initial field, the largest changes being in the adjustment of the pressure at low latitudes.

A complete program running the single level primitive equations model is provided in this chapter. Program (*SILEPE*) is relatively short and well documented. As previously mentioned, the shallow water model uses the same initialization program as the barotropic non-divergent model with the proper setting of coefficients *IOPT1* and *IOPT2*.

A twelve hour forecast of the wind field and geopotential height from the single level primitive equations model is presented as an illustration (Figure 11.3).

      program *SILEPE*

```
c
c
c     This program performs the single level primitive
c     equations forecast. It can also be used for iterative
c     initialization. The primitive equations use the zonal
c     and  meridional components of the wind  field and
c     the height field as predictive variables.
c     The advective terms are computed using a semi-
c     Lagrangian approach. Time integration is accomplished
c     through the Matsuno time integration scheme.
c     The model can be used with or without the terrain
c     field and its associated gradients. These fields
c     are assumed to be given. The model also allows for
c     closed or open boundaries. The domain invariants are
```

```
c     computed during each output time.
c
c     input      : (1)    Logical unit 21 contains the terrain field
c                         its associated zonal and meridional gradients.
c                         each array of data is stored in one record and
c                         are read in the zm, dhdx, dhdy order.
c                (2)     Logical unit 22 contains the initial zonal (u)
c                         and meridional(v) components of the wind
c                         field and height field (z). Each array is stored
c                         in one record in the u,v,z order.
c                (3)     Logical unit 23 contains information required
c                         by the program during execution. See
c                         comment cards below for details.
c
c     output     : (1)    Logical unit 24 contains the output fields.
c                         The first 3 records contain the initial u,v,z.
c                         Subsequent sets of 3 records contain the
c                         forecast u,v,z at 12 hourly intervals (or other
c                         intervals that has been set by the user.)
c                (2)     logical unit 6 is used for listing.
c
c
c
c     Variables input through unit 23.
c
c     variable name          description                    read in
c     -------------          -----------                    --------
c     ncase                  description of forecast          const
c     time                   starting time of forecast        const
c     ijk  (=0)              iterative initialization         const
c          (=1)              one level pe forecast
c     iterr  (=0)            no terrain effects included      const
c            (=1)            terrain included
c     ioutpt (=0)            no gain and shade done           const
c            (=1)            gain and shade done
c     nsbd  (=0)             v=0. at n-s boundary             const
c           (=1)             smoothing done at n-s boundary   const
c     alfa                   smoothing coeff (0. - 1.)        const
c     l,m,dphi               zonal dimension, meridional
c                            dimension and grid size in deg   const
c     slat,dt                southern most lat, time step     const
c     nohist                 history tape parameter           const
c                            (=999) no history tapes made
c
```

```
c
c      iterative initialization
c      ----------------------
c      niter                number of iterations                const
c      nout                 number of iter. before write        const
c                           (=1 for one time step)
c      itcnt                no. of history rec. to be read      const
c
c      forecast
c      --------
c      tfcst                length of forecast in hours         const
c      toint                output intervals                    const
c      tcnt                 no. of sets of records to be        const
c                           read
c
       parameter (lg = 51,mg = 21)
c
       common /a01/ l,     m,      l1,     m1,     m2,      g,
      1                    slat,   dphi
       common /a02/ jin,   jout,   itape,  jb,     jc,      jd
       common /a03/ ijk,   itcnt,  itime,  nstep,  iterr
       common /a04/ time,  dt,             nohist, ioutpt,  nsbd,   alfa
       common /a05/ tfcst, toint,  niter,  nout,   title(3)
       common /a06/ denom(6,mg), dx(mg),         cor(mg),  dy
       common /a07/ dx2iv(mg),   dxsqiv(mg),     dy2iv,    dysqiv
       common /a08/ zm(lg,mg),   dhdx(lg,mg),    dhdy(lg,mg)
       common /a09/ u(lg,mg),    v(lg,mg),       z(lg,mg)
       common /a10/ up(lg,mg),   vp(lg,mg),      zp(lg,mg)
       common /a11/ uo(lg,mg),   vo(lg,mg),      zo(lg,mg)
       common /a12/ a(lg,mg),    b(lg,mg),       c(lg,mg)
c
c      Open input and output files.
c
       open (21,file='terrain.out     ',status='old')
       open (22,file='infield_out.dat',status='old')
       open (23,file='tape23.dat',status='old')
       open (24,file='tape24.dat',status='new')
c
c      Initialize variables in the common block.
c
       call ZEROS
c
c      Call subroutine CONSTZ to define constant
```

```
c      parameters and constants required by the program
c
       call CONST2
c
       write(6,1000) time
1000   format(5x,'starting time of forecast is ',f5.0)
1001   format(30x,'iterative initialization')
1002   format(10x,'number of iterations =',i3,2x,'no of iterations ',
     & 'between outputs =',i2,2x,'no of history records to be read =',
     & i3)
1003   format(30x,' pe. forecast ')
1004   format(10x,'length of forecast in hrs =',f6.3,3x,
     & ' output interval in hrs =',f6.2,3x,' no of history records,
     & to be read =',i5)
1005   format(10x,'l=',i5,2x,'m=',i5,2x,'slat=',f5.1,2x,
     & 'dphi=',f6.3,2x,'dt=',f5.1)
c
       if (ijk.eq.0) then
       write(6,1001)
       write(6,1002) niter, nout, itcnt
       else
  45   continue
       write(6,1003)
       write(6,1004) tfcst, toint, itcnt
       endif
       write(6,1005) l, m, slat, dphi, dt
c
c      Call subroutine INDATA to  read in the
c      terrain and initial fields.
c
       call INDATA
c
c      Call subroutine INVART to compute the
c      closed domain invariants.
c
       call INVART
c
       do 11100 j = 1, m,m1
       do 11100 i = 1, l
           uo(i,j)   = u(i,j)
           vo(i,j)   = v(i,j)
11100      zo(i,j)   = z(i,j)
       jin          = ifix((toint*3600.)/dt)*itcnt
```

```
          if (itcnt.eq.1) jin = 0
          if (ijk.eq.0  ) then
          jint        = nout*nstep
          if (jint.eq.0 ) jint = 1
          jout        = jin + jint
          stop        = niter/nout
          else
          jint        = toint*3600./abs(dt)
          if (jint.lt.1) jint = 1
          jout        = jin+jint
          jstop       = tfcst/toint+0.5
          endif
c
c     The 'do loop 11102' controls the output of the
c     forecast products. During each loop
c     subroutine 'FCST' is first called to perform
c     the time integration until the output time
c     is reached. The forecast products are then written
c     to the output file if required. Next, subroutine
c     'INVART' is called to compute the closed domain
c     invariants of the model.
c
          do 11102 kk = 1, jstop
             call FCST
             itape     = 24
             if (nohist.ne.999) then
             write (itape,878)  ((u(i,j),i=1,l-6),j=1,m)
             write (itape,878)  ((v(i,j),i=1,l-6),j=1,m)
             write (itape,878)  ((z(i,j),i=1,l-6),j=1,m)
 878      format(8f10.4)
             endif
             call INVART
             write(6,1006) kk
1006      format(2x,'after invart',5x,'output step number = ',i5)
             jin       = jout
             jout      = jin+jint
11102     continue
          end
```

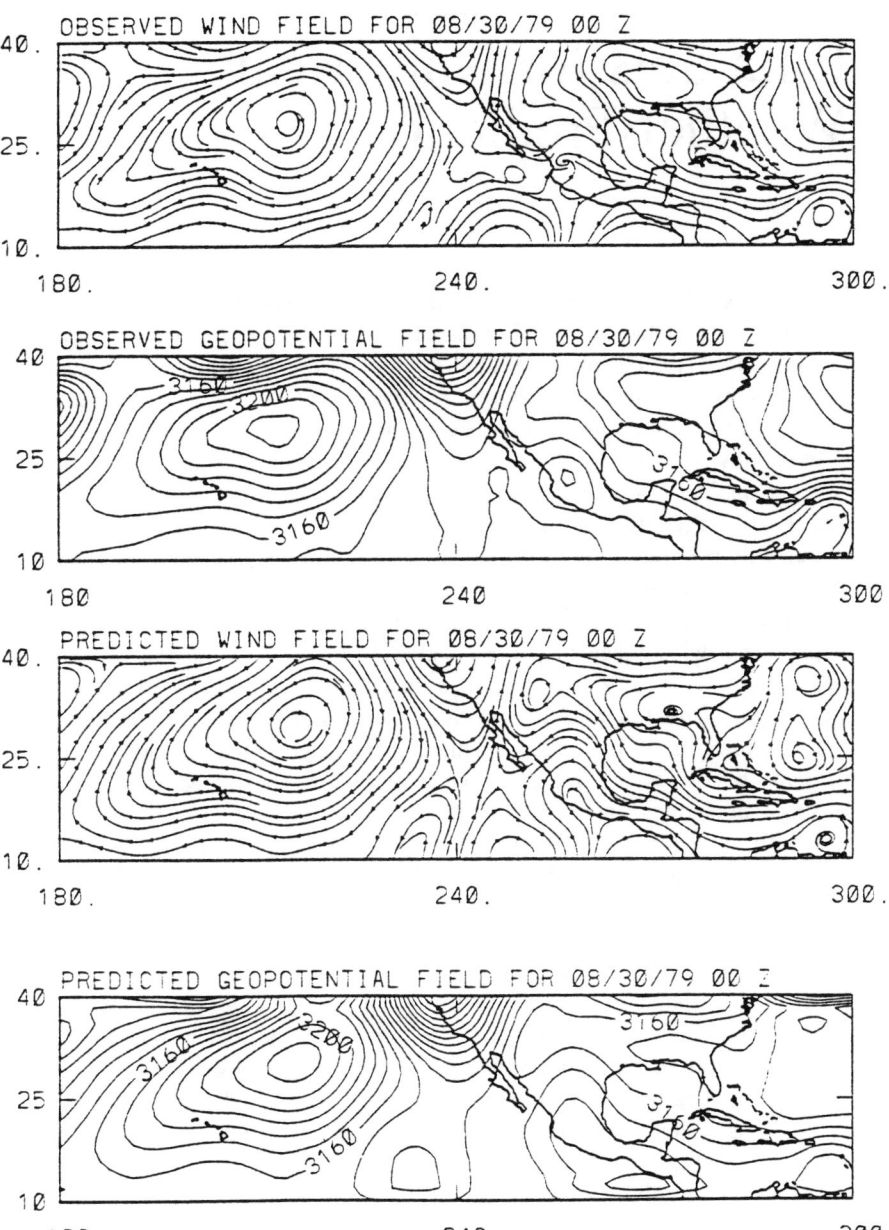

Figure 11.3: Twelve hour forecast of wind and geopotential fields at 700
mb valid August 30, 1979 at 00 GMT.

12

Data Sets for Numerical Weather Prediction

During the last decade, satellite based sensors have provided scientists with objective and reliable global data sets which have revolutionized their ability to observe the earth-atmosphere evolution. Remote sensing provides a synoptic observation of the earth and appears in most cases to be more consistent and reliable than ground based observations which suffer from an uneven spatial distribution and a varying quality. Although conventional station data cannot provide a global coverage, they still constitute the ground truth for satellite instruments calibration and are, at the present time, a necessary element for numerical weather prediction.

Space based sensors measure only radiation which is then analyzed to provide a wide range of geophysical parameters. Satellite instruments measure the intensity of the electromagnetic radiation at different wavelengths. This recorded information is next transmitted and converted into geophysical data using mathematical algorithms. Among the elements of the earth's radiative budget that are considered in remote sensing calculations, one can list the direct solar radiance reflected from the earth and the radiation emitted or scattered by the atmosphere downward to the earth and then reflected upward. The amount of radiation emitted from the ground and that emitted or scattered by the atmosphere towards the sensors constitute also important components (Gurney et al., 1993). The observing instrument aboard the satellite is designed to respond specifically to the electromagnetic properties of the observed material. For example the ultra-violet, visible and near infrared channels are used to measure the reflected solar energy so that the reflectance properties of the substance are emphasized. The emittance of the body is better examined by measuring the emitted energy in the microwave and infrared channels.

Spaceborne observations have surely resulted in a step forward concerning the understanding of the natural variability of the earth-atmosphere-ocean system; nevertheless, at the present state of the art, only few numerical models are capable of directly assimilating them. Satellite data provide a wide range of parameters such as atmospheric ozone, vegetation and ice cover, cloud motion, temperature and humidity profiles, sea surface temperature, etc., which constitute a useful background for numerical weather prediction.

On the other hand, in addition to the conventional data base provided by the world weather watch (WWW) and which include radiosonde-rawinsonde, pilots balloons and surface observations, numerical weather prediction modelers should be aware of several other data sets that are currently part of the operational stream. These include commercial aircraft wind reports from wide bodied jets equipped with inertial navigation systems, marine surface observations from ships of opportunity and data from moored and drifting buoys. Some other data sets such as fractional cloud cover and soil moisture are provided by the models themselves.

This chapter describes some algorithms used to retrieve meteorological parameters needed for numerical weather prediction from satellite data. It also presents an example of model generated data.

1. Rainfall Distribution from Outgoing Longwave Radiation

In meteorology, many diagnostic and prognostic problems require the knowledge of rainfall fields. For example, in numerical weather prediction rainfall rates are useful for the specification of the diabatic fields and the physical initialization. They also serve as a forecast verifying analysis. Availability of global precipitation fields presents some difficulty, especially over oceanic regions where observations are virtually inexistent. The use of satellite data for estimating the rainfall pattern has thus become necessary for numerical weather prediction modelers.

A simple algorithm constructing a rainfall field from satellite infrared brightness and raingauge data is presented in this section. The method makes use of a statistical multiple regression and an objective analysis scheme. The ultimate output of this technique is a rainfall field which takes into account both available raingauge data and satellite brightness data sets for each day.

Given times series of rainfall data, R, at N observing sites, the first step consists of interpolating the satellite infra-red brightness, T_{iB}, and its time rate of change onto the observing locations. This procedure provides time series of three collocated variables, $R(t)$, $T_{iB}(t)$ and $\frac{\partial T_{iB}}{\partial t}(t)$. Satellite data are provided by the National Oceanic and Atmospheric Administration (NOAA) polar orbiting satellite TIROS N which has been operational since 1979. This satellite is sun-synchronous with an equator crossing around 3:00 a.m. and 3:00 p.m. local time. Therefore, the collocated variables require a systematic synchronization. For example, raingauge rainfall are totals representing a 24-hour period. Therefore, mean values of brightness data and their time tendency need to be obtained over the same period. This is done by an interpolation of satellite data from 3 a.m. to 3 p.m. data sets to a uniform Greenwich base at 0000 Z.

The next step consists of performing a multiple linear regression between the rainfall and the three collocated variables to provide a least square fit of the form,

$$R = a\,T_{iB} + b\,\frac{\partial T_{iB}}{\partial t} + c \tag{12.1}$$

Satellite data corresponding to no rain conditions are excluded from the regression analysis.

Using data over the Asian monsoon region (30S - 30N and 60E - 165E) during the summer of 1979, the regression coefficients were estimated at

$$a = -\,0.1824 \quad b = -\,0.3848 \quad c = 52.41 \tag{12.2}$$

It should be noted that these coefficients may be useful only with the use of TIROS N data and for a 24 hour total rainfall. The negative coefficient of T_{iB} suggests that taller clouds have colder black body temperatures and would contribute to large rainfall. On the other hand, growing clouds would exhibit a negative time tendency for T_{iB} and would then contribute to larger rainfall rates.

The second part of this scheme uses an objective analysis scheme to obtain an adequate rainfall distribution on a regularly spaced grid array. Given the satellite data over a grid point at a particular day, the statistical regression coefficients are used to provide a first rainfall guess field of rainfall. Raingauge data are used as the observational data base and an objective analysis using the successive correction method, as described in Chapter 5, is performed. The final rainfall field is obtained after four scans are performed reducing the optimal radius of influence from $4\Delta x$ to Δx, where Δx is the grid size.

Rainfall estimates obtained through this scheme show a reasonable distribution and are in good agreement with raingauge data over land. Over data void oceans, the analysis resembles the first guess field.

The code carrying this analysis is not provided in this workbook. However, the least square approach used in the multiple linear regression is described in Chapter 3 and the successive correction method is presented in Chapter 5. A flow chart showing the computational aspect of this scheme is presented in Figure 12.1. There exist several other statistical algorithms for the derivation of rain rates from the outgoing longwave radiation (Arkin, 1994). A sample map of outgoing longwave radiation over the monsoon domain is presented in Figure 12.2. Figure 12.3 shows the rainfall derived from the OLR field.

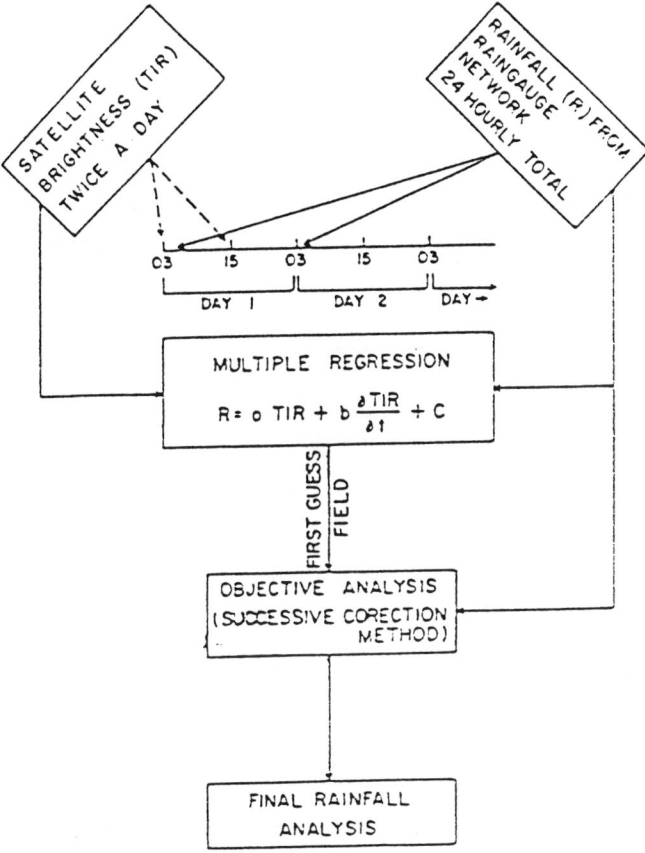

Figure 12.1: Algorithm for the computation of rainfall from satellite infrared brightness.

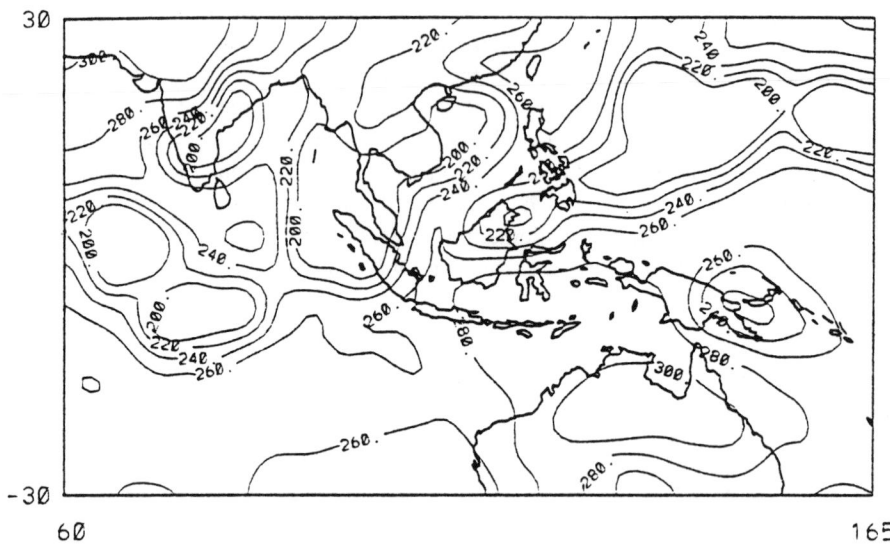

Figure 12.2: Outgoing longwave radiation (OLR) over the tropics (watts/m²), for October 6, 1991. Contours interval = 20 watts m⁻².

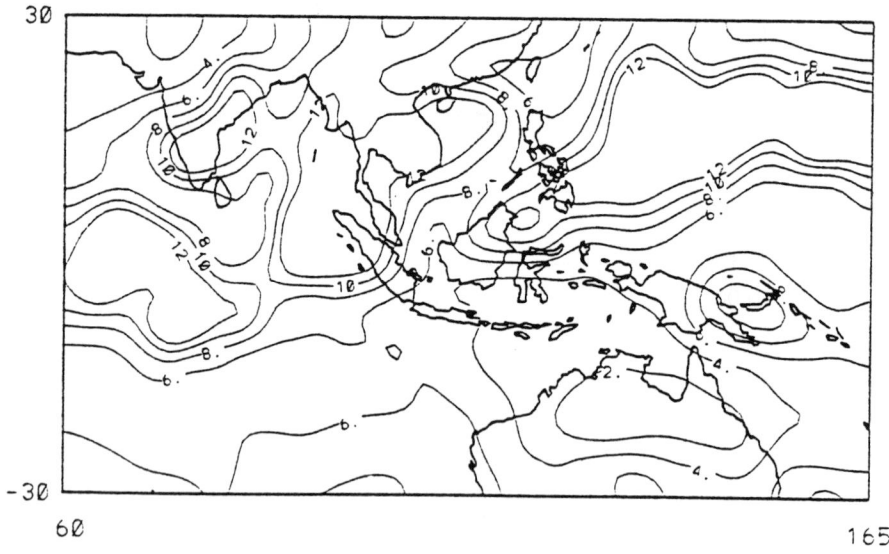

Figure 12.3: Rainfall distribution determined from an OLR rain rate algorithm (mm/day), for October 6, 1991. Contour interval = 2 mm day⁻¹.

2. SSM/I Based Rain Rates, Wind Speed and Total Precipitable Water

Rainfall rates can also be obtained from algorithms such as those developed by Olson et al. (1990). These are based on the multiple regression method and regress the rainfall against the brightness temperature (TB) obtained from different microwave sensors of the special sensor microwave/imager (SSM/I) on board the defense meteorological satellite program (DMSP) satellites. The coefficients of regression presented in this chapter were obtained by regressing observed rain rates (from raingauge measurements) against colocated microwave measures of the brightness temperature in various channels.

The following are some useful algorithms which provide these fields as a function of the SSM/I data sets at the different frequencies.

2.1 Rain Rates

2.1.1 Rainfall Over Land Using the 85 GHz Channel

$$R = \exp(3.29716 - 0.01290\ TB^{85V} + 0.0087\ TB^{85H}) - 8.0$$

$$(12.3)$$

Here R denotes the rainfall rate in millimeters per hour and TB^{85} denotes the brightness temperature data at 85 GHz. The superscripts V and H refer to vertical and horizontal polarization, respectively.

2.1.2 Rainfall Over Ocean Using the 85 GHz Channel

$$R = \exp(3.06231 - 0.0056036\ TB^{85V} + 0.0029478\ TB^{85H}$$
$$- 0.0018119\ TB^{37V} - 0.00750\ TB^{22V} + 0.0097550\ TB^{19V})$$
$$- 8.0 \qquad (12.4)$$

In case the 85 GHz channels were unusable the regressions were built using the 19 and 37 GHz channels.

2.1.3 Rainfall Over Land Using the 37 GHz Channel

$$R = \exp(-17.76849 - 0.09612\ TB^{37V} + 0.15678\ TB^{19V}) - 1.0$$

$$(12.5)$$

2.1.4 Rainfall Over Ocean Using the 37 GHz Channel

$$R = \exp(5.10196 - 0.05378 \ TB^{37V} + 0.02766 \ TB^{37H}$$
$$+ 0.01373 \ TB^{19V} - 2.0 \tag{12.6}$$

2.2 Wind Speed Over Oceans

The regression equation for the wind speed over oceans is based on Halpern et al. (1993). The coefficients are derived by calibrating against marine ship observations.

$$WS = 147.9 + 1.0969 \times TB^{19V} - 0.455 \times TB^{22V}$$
$$- 1.7600 \times TB^{37V} + 0.786 \times T^{37H} \tag{12.7}$$

2.3 Total Precipitable Water

The following equation for the total precipitable water is also obtained from the regression approach. The radiosonde based measures of total precipitable water and colocated SSM/I brightness temperatures are used to obtain the coefficients. This is based on the study of Filiberti et al. (1993).

$$PW = 20.75 - 2.582 \ \log_e(280. - TB^{19H}) - 0.3919 \ \log_e(280.$$
$$- TB^{19V}) - 3.610 \ \log_e(280. - TB^{22V}) + 2.729 \ \log_e(280.$$
$$- TB^{37H}) - 0.5118 \ \log_e(280. - TB^{37V}) \tag{12.8}$$

Brightness temperature corrections were applied to eliminate the weak biases observed between the temperature measured by SSM/I and those computed by the European Center for Medium Range Weather Forecast (ECMWF) model. The corrected brightness temperatures T' were expressed as linear functions of the measured brightness T as

$$T' = aT + b \tag{12.9}$$

where the coefficients a and b are defined in Table 12.1.

Table 12.1 Correction Coefficients

	a	b
19H	0.948	8.3
19V	0.927	12.3
22V	0.910	17.4

The validation of the 37 GHz channel correction was not possible with the same method because this channel is too sensitive to water content. The data sets provided by the above algorithms serve a very useful purpose in providing coverage over data sparse tropical oceans. The full potential of these data for numerical weather prediction has not been fully realized at present.

Most of these algorithms are easy to program and are therefore illustrated in one single driver. Program (*SSMI*) makes use of a sample 10 x 10 input array of SSM/I brightness temperatures to compute the rainfall, precipitable water and wind speed. Rainfall and total precipitable water derived from SSM/I are shown in Figures 12.4 and 12.5, respectively.

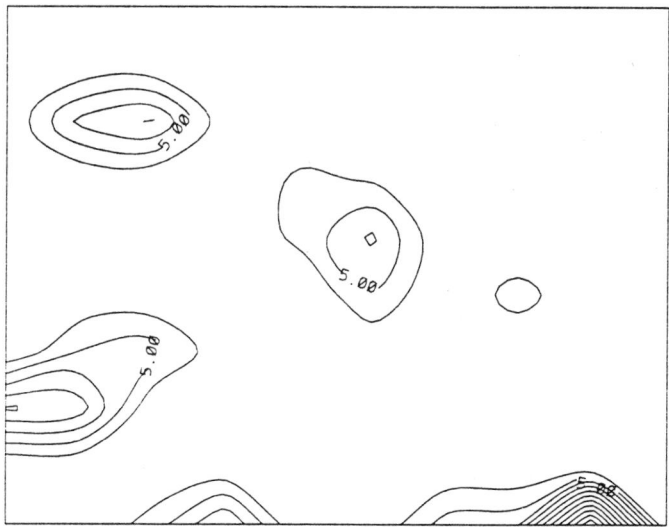

Figure 12.4: Rainfall rate distribution inferred from SSM/I data algorithm (mm/day), for October 6, 1991. Contour interval = 2.5 mm day^{-1}.

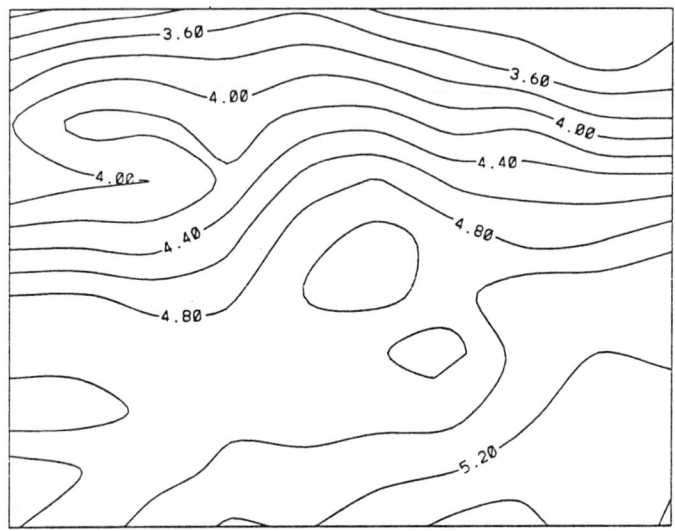

Figure 12.5: Total precipitable water (kg/m^2) determined from
            SSM/I algorithm, for October 6, 1991. Contour interval
            = 0.2 kgm^{-2}.

        program *SSMI*

```
c
c    This program uses algorithms described in Chapter 12
c    to estimate the rainfall, wind and precipitable water
c    from SSM/I input data.
c    100 sample ssm/i brightness temperatures are given on
c    a (10,10) grid. All 7 channels are provided. The domain
c    considered for this input extends from 157.5 east to
c    167.625 east and from 0.561 north to 10.654 north.
c
c    Parameters :
c
c    i85chk      : is a flag for using the 85 Ghz. channel
c    isfcl       : is a flag for land-sea mask
c    isfcl = 5   : (ocean)
c    isfcl = 1   : (vegetated land)
c    isfcl = 0   : (land)
c    tb7         : is an input array containing ssm/i data
c                     for the 7 channels, on a 10x10 matrix.
c
```

```
c
        logical i85chk
        real tb7(7,10,10),rws(10,10,3)
c
c    Open the input and output files
c
        open (10,file='ssmitb7  ',status='old')
        open (11,file='rws10.dat',status='unknown')
c
        i85chk    = .true.
c
c    Note that all points in this domain are
c    located over ocean
c
        isfcl     = 5
c
c  Read input data
c
        read(10,'(7f7.2)') tb7
c
c    Perform calculations one grid point at a time.
c
        do 200 j  = 1, 10
        do 200 i  = 1, 10
c
c    In the following variable names, the number
c    refers to the channel and the v and h refers
c    to the polarization.
c
        rh85      = tb7(1,i,j)
        rv85      = tb7(2,i,j)
        rh37      = tb7(3,i,j)
        rv37      = tb7(4,i,j)
        rv22      = tb7(5,i,j)
        rh19      = tb7(6,i,j)
        rv19      = tb7(7,i,j)
c
c    NOTE : The screening logic for the ocean
c    and land algorithms follows the reports:
c    'Recommended Algorithms for the Retrieval
c    of Rainfall Rates in the Tropics Using the
c    SSM/I ( DMSP-8) '  W.S. Olson, et al. (1990).
c    And 'DMSP Special Sensor Microwave/Imager
```

```
c     Calibration/Validation' Coordinated by
c     J.P Hollinger.
c
c
c     Initialize the precipitation field.
c     pcp is given in (mm/hr)
c

      pcp        = 0.0
c
c     Ocean Algorithm
c
               if (isfcl.eq.5 ) then

         if((-11.7939-.02727*rv37+.09920*rh37).gt.0.0) then
c
c     Use algorithm defined for the 85 Ghz. channel
c
      if(i85chk) then
c
c     Hughes' negative polarization test for bad data.
c
      if((rv85-rh85).lt.-2.) then
      pcp        = 0.0
      else
      pcp        = exp ( 3.06231 - .0056036*rv85 + .0029478*rh85
     &             - .0018119*rv37 - .00750*rv22 + .009755*rv19)
     &             - 8.0
      endif
               else
c
c     Use algorithm defined for the 37 Ghz. channel
c
      pcp        = exp (5.10196 - .05378*rv37 + .02766*rh37
     &             + .01373*rv19) - 2.0
      endif
      endif
               endif
c
c     Land and vegetated-land Algorithms
c
      if (isfcl.eq.0.or.isfcl.eq.1) then
c
c     Use algorithm defined for the 85 Ghz. channel
```

```
c
        if(i85chk) then
c
c    Hughes' negative polarization test for bad data.
c
        if ((rv85-rh85).lt.-2.) then
        pcp        = 0.0
        else
c
c    Check for measurments over vegetated land.
c
        if ((rv22-rv19.le.4.0.and.(rv19+rv37)/2.-(rh19+rh37)/2.
     +      .le.4.0.and.rv85-rv37.lt.-1.0.and.rv19.gt.268.) .or.
c
c  Check for measurments over bare land.
c
     +      (rv22-rv19.le.4.0.and.(rv19+rv37)/2.-(rh19+rh37)/2.
     +      .gt.4.0.and.rv37-rv19.lt.-3.0.and.rv85-rv37.lt.-5.0
     +      .and.rh85-rh37.lt.-4.1.and.rv19.gt.268.))
c
     +      pcp        = exp(3.29716 - .01290*rv85 + .00877*rh85)
     +                   - 8.0
        endif
        else
c
c    Calculations in the case the 85 Ghz. channel is not available.
c
        if((rv22-rv19.le.4.0.and.(rv19+rv37)/2.-(rh19+rh37)/2.
     +      .le.4.0.and.rv37-rv19.lt.-6.4.and.rv19.gt.268.) .or.
     +      (rv22-rv19.le.4.0.and.(rv19+rv37)/2.-(rh19+rh37)/2.
     +      .gt.4.0.and.rv37-rv19.lt.-6.4.and.rv19.gt.268))
     +      pcp        = exp(-17.76849 - .09612*rv37 + .15678*rv19)
     +                   - 1.0
        endif
        endif
c
c    Set the precipitation to zero in case it is negative,
c    and convert it to mm./day.
c
        if (pcp.lt.0.0) pcp  =  0.0

        pcp        = pcp * 24.0
```

```
c
c     Compute wind speed (m/s)
c
      spd      = 0.0
      if (pcp.le.0.0.and.isfcl.eq.5)   ! why
   &     spd = 147.9 + 1.0969*rv19 - 0.455*rv22
   &             - 1.7600*rv37 + 0.786*rh37
      if (spd.lt.0.0) spd = 0.0
c
c     Compute integrated water vapor (g/cm**2)
c
      wtvap     = 0.0
      rh19c     = 0.948*rh19+8.3
      rv19c     = 0.927*rv19+12.3
      rv22c     = 0.91 *rv22+17.4
      if (rh19c.lt.280..and.rv19c.lt.280..and.rv22c.lt.280..and.
   &     rh37.lt.280..and.rv37.lt.280..and.isfcl.eq.5)
   &     wtvap = 20.75 - 2.582*log(280.-rh19c) - 0.3919*log(280.-rv19
   &                 - 3.6100*log(280.-rv22c) + 2.729*log(280.-rh37)
   &                 - 0.5118*log(280.-rv37)
c
c     Save calculated values in rws array
c     and continue for next grid point.
c
      rws(i,j,1) = pcp
      rws(i,j,2) = wtvap
      rws(i,j,3) = spd
c
 200     continue
c
c     Display output
c
 1001    format(/,15x,'ssm/i rainfall field ',/)
 1002    format(/,15x,'ssm/i precipitable water field ',/)
 1003    format(/,15x,'ssm/i wind speed  field ',/)
      write (6,1001)
      write(6,'(10f7.2)') ((rws(i,j,1),i=1,10),j=1,10)
      write (6,1002)
      write(6,'(10f7.2)') ((rws(i,j,2),i=1,10),j=1,10)
      write (6,1003)
      write(6,'(10f7.2)') ((rws(i,j,3),i=1,10),j=1,10)
      stop
      end
```

3. Normalized Difference Vegetation Index

A considerable interest in remotely sensing the biophysical characteristics of the land surface has emerged during the past few years as a contribution to the monitoring of the global climate change. The role of the vegetation on the control of the global water cycle has led to many investigations aiming at functional relationships between the spectral reflectance of various vegetation species and their biological and physiological characteristics. Among many spectral indices that describe the vegetation density, the normalized difference vegetation index (NDVI) seems to be the most popular. The NDVI is defined on the basis of the differential reflection characteristic of the green fraction of the vegetation in the visible and infrared regions of the solar spectrum (Tucker and Miller, 1977). This vegetation index is obtained as a combination of radiances measured by the advanced very high resolution radiometer (AVHRR) aboard the National Oceanic and Atmospheric Administration NOAA-8 and NOAA-9 polar orbiting satellites and is defined as

$$\text{NDVI} = \frac{R_{ch2} - R_{ch1}}{R_{ch2} + R_{ch1}} \tag{12.10}$$

where R_{ch1} and R_{ch2} are the reflectances in the visible (0.58 to 0.68 μm) and near infrared (0.725 to 1.1 μm) channels, respectively. NDVI has been shown to be a good indicator of the green vegetation density (Tucker et al., 1981; Curran, 1980). Further studies have related the NDVI to the amount of photosynthetically active radiation (PAR) absorbed by the green fraction of the canopy and have demonstrated that NDVI is indeed a measure of the photosynthetic capacity of the canopy and bulk stomatal conductance to transfer of water vapor (Sellers, 1985; Tucker and Sellers, 1986). The NDVI is used extensively in biophysically based land surface models to determine vegetation characteristics.

4. Fractional Cloud Cover

Fractional cloud cover as obtained by (9.27), (9.28) and (9.29) in the radiative transfer scheme are presented in Figure 12.6 as an example of model generated data set.

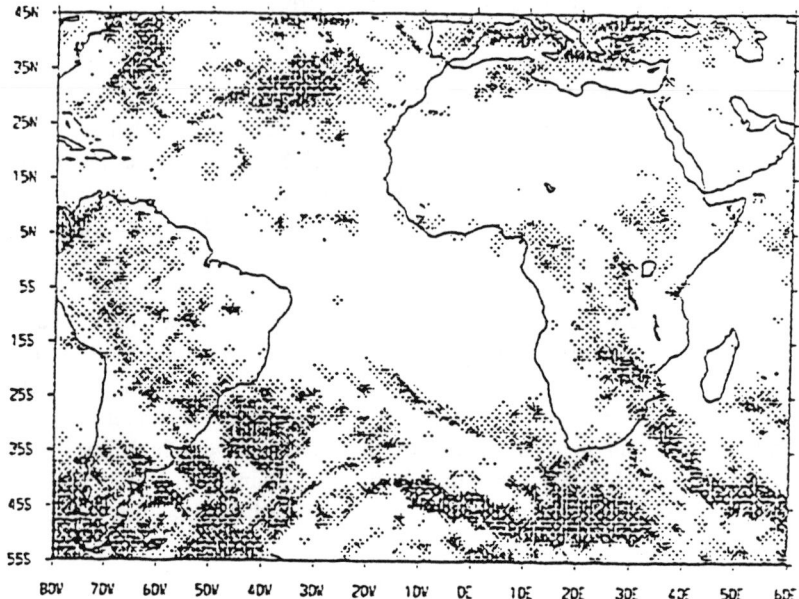

Figure 12.6: Typical model based total cloud fraction distribution over
             the tropics (high, medium, low). Shading interval for
             fractional cloudiness = 20%.

13

Model Output Diagnostics

A number of diagnostic computations can be made to examine model output and thus the model performance. Such computations are not standard and may differ according to the objective pursued. For example, it is possible to evaluate energy quantities and energy transformations using the predicted values. If a meteorological phenomena such as a storm forms during the period of model integration, the departure of the kinetic energy from a zonal mean is expected to increase, and the budget of this quantity would be quite revealing. The diagnostics can be used either to reveal the processes involved in the development of a meteorological phenomenon or to explore the model deficiencies in case the forecast was not successful.

There are many other areas where model output diagnostics are useful, for example the monitoring of water and energy budgets. This chapter presents some important aspects of energy and energy transformations that can be used to carry out diagnostics. It also introduces the reader to the notion of trajectory. Many computational algorithms are provided but are not all used in the one example given, (*DIAGNOS*). These subprograms are however extensively documented and fully tested. A four-dimensional trajectory computation program, (*TRAJECT*), is also provided.

1. Energy and Energy Transformation Terms

A standard formulation of the energy equations are followed in this section. Before defining these equations some mathematical operators need to be introduced. The zonal average operator is defined as

$$[(\quad)] = \frac{1}{\lambda_e - \lambda_w} \int_{\lambda_w}^{\lambda_e} (\quad) \, d\lambda \qquad (13.1)$$

where λ_e and λ_w represent the eastern and western longitudes of the domain, respectively. The area average operator is also needed and is given by

$$\overline{(\quad)} = \frac{1}{\sin\phi_n - \sin\phi_s} \int_{\phi_s}^{\phi_n} [(\quad)] \cos\phi \, d\phi \qquad (13.2)$$

where ϕ_s and ϕ_n denote the southern and northern latitudes, respectively. These two operations are performed by subroutine *AVERAG1*.

Furthermore, the deviations from the area average and from the zonal average are calculated as

$$(\quad)'' = (\quad) - \overline{(\quad)} \qquad (13.3)$$

and

$$(\quad)' = (\quad) - [(\quad)] \qquad (13.4)$$

Finally the departure of the zonal average from the area average is estimated as

$$(\quad)^* = [(\quad)] - \overline{(\quad)} \qquad (13.5)$$

These calculations can be carried out using *DEVIAT*, while subroutine *INTEGR* is used to perform the vertical integration.

1.1 Energy Terms

The zonal available potential energy is expressed as

$$AZ = \int_{100}^{p_s} \frac{\overline{T^{*2}}}{2\sigma} \, dp \qquad (13.6)$$

where

$$\sigma = g \left\{ \frac{R_d \, [\overline{T}]}{c_p p} - \frac{\partial [\overline{T}]}{\partial p} \right\} \qquad (13.7)$$

defines the stability of the atmosphere. The terms g and R_d represent the acceleration due to gravity and the gas constant for dry air, respectively. The atmosphere is assumed to be bounded by the surface pressure, p_s, and 100 mb aloft. The temperature deviation from its zonal average is used to define the eddy available potential energy as

$$AE = \int_{100}^{P_S} \overline{\frac{T'^2}{2\sigma}} \, dp \qquad (13.8)$$

On the other hand, the zonal kinetic energy is written as

$$KZ = \frac{1}{2g} \int_{100}^{P_S} \overline{([u]^2 + [v]^2)} \, dp \qquad (13.9)$$

where u and v denote the horizontal wind components, and the eddy kinetic energy is defined in terms of deviations from the zonal average.

$$KE = \frac{1}{2g} \int_{100}^{P_S} \overline{(u'^2 + v'^2)} \, dp \qquad (13.10)$$

Subroutine *APEZANDE* computes both the zonal available potential energy and the eddy available potential energy as defined in (13.6) and (13.8). The zonal kinetic energy and its eddy component are carried out by *KEZANDE*. Subroutine *STASTB* is provided for the calculation of the stability parameter and *AVMERID* for the calculation of the meridional mean.

1.2 Energy Transformation Terms

The first term in the series of these important energy transformations describes the conversion of zonal available potential energy into eddy available potential energy,

$$CA = -\int_{100}^{P_S} \left[\frac{1}{\sigma} \overline{v'T' \frac{\partial T}{a\partial \phi}}^* + \frac{1}{\sigma} \overline{\omega'T' \frac{\partial T}{\partial p}}^* \right] dp \qquad (13.11)$$

The transformation of zonal kinetic energy into eddy kinetic energy is expressed as

$$
CK = \frac{1}{g} \int_{100}^{p_S} \left[\cos\phi \, u'v' \, \frac{\partial}{a\partial\phi} \left[\frac{[u]}{\cos\phi} \right] \right] dp + \frac{1}{g} \int_{100}^{p_S} \left[\overline{v'^2 \, \frac{\partial[v]}{a\partial\phi}} \right] dp
$$

$$
+ \frac{1}{g} \int_{100}^{p_S} \overline{\frac{\tan\phi}{a} \, u'^2[v]} \, dp + \frac{1}{g} \int_{100}^{p_S} \left[\overline{\omega'u' \, \frac{\partial[u]}{\partial p}} \right] dp
$$

$$
+ \frac{1}{g} \int_{100}^{p_S} \left[\overline{\omega'v' \, \frac{\partial[v]}{\partial p}} \right] dp \tag{13.12}
$$

These two terms are computed simultaneously in subroutine **CKCATERM**. Subprograms **DERIVE** and **PRESDE** are provided for calculations of the meridional and vertical derivative, respectively.

The conversion of eddy available potential energy into eddy kinetic energy is done following

$$
CE = -\frac{1}{g} \int_{100}^{p_S} \frac{R}{p} \, \overline{\omega'T'} \, dp \tag{13.13}
$$

and the zonal available potential energy is transformed into zonal kinetic energy as

$$
CZ = -\frac{1}{g} \int_{100}^{p_S} \frac{R}{p} \, \overline{\omega^* T^*} \, dp \tag{13.14}
$$

Subroutine **CECZTERM** performs these calculations.

1.3 Energy Generation Terms

The term representing the generation of the available potential energy by differential heating can be divided into a zonal part and an eddy part. The zonal part, G_z, is represented by

$$G_z = \frac{R_d}{C_p} \oint \frac{[\theta]^* \ [Q]^*}{p(-\frac{\partial\bar\theta}{\partial p})} \ dm \qquad (13.15)$$

The eddy part, G_E, is represented by

$$G_E = \frac{R_d}{C_p} \oint \frac{(\theta' \ Q')}{p(-\frac{\partial\bar\theta}{\partial p})} \ dm \qquad (13.16)$$

where $Q = C_p \frac{dT}{dt} - \alpha\omega$ is the heating rate per unit mass and $dm = \rho$ dx dy dz represents a mass element.

An obvious drawback of the above formulation arises from the boundary flux terms that need to be invoked for completeness. In this study the time history of these energy quantities and their transformations is examined. It is assumed that the domain is sufficiently large so that the internal processes would be revealed from the aforementioned analysis. The computation of these energy generation terms is done by subroutine *GZGETERM*. Program (*DIAGNOS*) is a driver for some of the above mentioned subroutines.

Figure 13.1(a-d) shows respectively the time history of the zonal available potential energy, the eddy available potential energy, the eddy kinetic energy and the zonal kinetic energy during a formation of a hurricane.

The energetics presented in this section are integrals over an Atlantic domain covering 310.5E to 330.5E and 0.5N to 20.5N. The units of the energy quantities are expressed in x 10^5 Jm^{-2}. The energy transformation quantities are expressed in Wm^{-2}.

The time history of the salient energy transformations and of its generation are shown in Figure 13.2(a-f). Here we show respectively the conversions from zonal available to the eddy available potential energy, from zonally available potential energy to the zonal kinetic energy, from eddy available potential to the eddy kinetic energy, and from the zonal kinetic to the eddy kinetic energy and the generation terms.

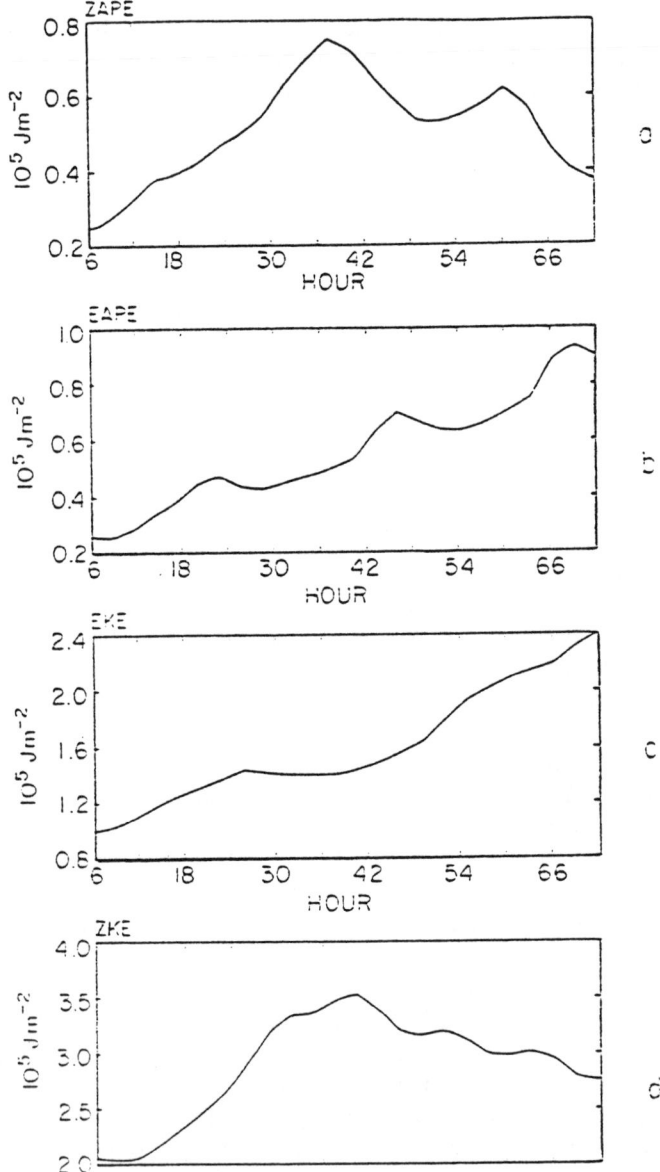

Figure 13.1: Time evolution of (a) zonal available potential energy, (b)
              eddy available potential energy, (c) eddy kinetic energy,
              and (d) zonal kinetic energy from 06UTC, August 30,
              1979 through 00UTC, September 2, 1979 (units: $10^5 Jm^{-2}$).

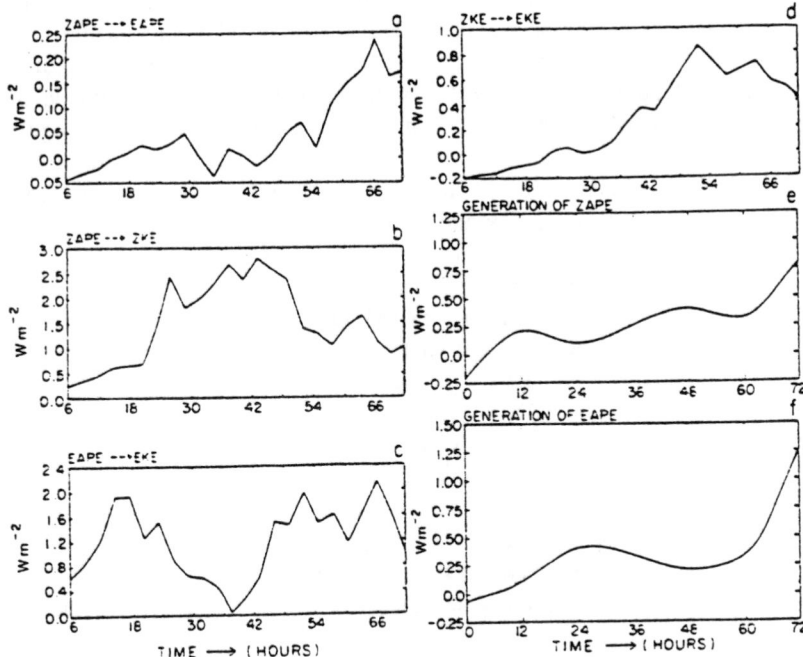

Figure 13.2: Time evolution of the energy conversions and energy generations from 06Z, August 30, 1979 through 00Z, September 2, 1979. (a) conversion of zonal available potential energy to eddy available potential energy, (b) conversion of zonal available potential energy to zonal kinetic energy, (c) conversion of eddy available potential energy to eddy kinetic energy, (d) conversion of zonal kinetic energy to eddy kinetic energy, (e) generation of zonal available potential energy, and (f) generation of eddy available potential energy (units: Wm^{-2}).

2. On the Computation of Four-Dimensional Trajectories

2.1 Introduction

Trajectory calculations are required in many geophysical problems where it is important to trace in time the path of the parcels carrying a certain tracer or property. Trajectory programs are based on the principle that at any point on the trajectory path, the three-dimensional displacement is produced by the components of the winds u, v and w at that point. The wind components are not available at all points in the three dimensional space. They are normally available only at some coarse but regular grid points. Therefore, the wind components at any point on the path or trajectory are obtained by interpolation from wind data at regular three dimensional grid. The displacements at any point along the trajectory are obtained for a short period of time using Newton's Law. The displacements along x, y and p axes are described as

$$
\begin{aligned}
Dx &= u.Dt + 0.5.ax.Dt.Dt \\
    &= 0.5(u + (u+ax.Dt)).Dt \\
    &= 0.5(u + u')Dt
\end{aligned}
\tag{13.17}
$$

$$
\begin{aligned}
Dy &= v.Dt + 0.5.ay.Dt.Dt \\
    &= 0.5(v + (v+ay.Dt)).Dt \\
    &= 0.5(v + v')Dt
\end{aligned}
\tag{13.18}
$$

$$
\begin{aligned}
Dp &= w.Dt + 0.5.ap.Dt.Dt \\
    &= 0.5(w + (w+ap.Dt)).Dt \\
    &= 0.5(w + w')Dt
\end{aligned}
\tag{13.19}
$$

where ax, ay, and ap are the accelerations along x, y, and p axes, respectively. The prime terms u', v', and w' represent velocities components at the end point of the trajectory segment. Initially the first guess displacements are computed without acceleration. The end point wind components u', v', and w' are interpolated and used in (13.17) to (13.19) to estimate the displacement. The values of u', v', and w' can be improved by iterating this procedure. From the starting point of trajectory, the next point is obtained by displacing by Dx, Dy, and Dp along x, y, p axes. This procedure is followed throughout the entire path of the trajectory.

2.2 Trajectory Calculation

A program for the computation of forward and backward trajectories is provided (*TRAJECT*). The input wind components are read sequentially in chronological order depending on the type of trajectories desired. Data are next interpolated in time to a 10 minute temporal resolution which is the time step required for the computation. The starting and ending positions for the required trajectories are defined in terms of degrees of latitude, longitude and pressure. Subroutine *LOCATE* converts these positions into the nearest i, j, k indexes to the south, west and lower pressure data points in the domain. It also computes the distances of the trajectory point from this data point. The three wind components at the starting point of the trajectories are interpolated from the grid data, using a trilinear interpolation scheme *INTRPL*. Subroutine *DISPLACE* computes the displacement of the parcels along the trajectory and determines the next point.

An example of trajectory calculation is shown in Figure (13.3).

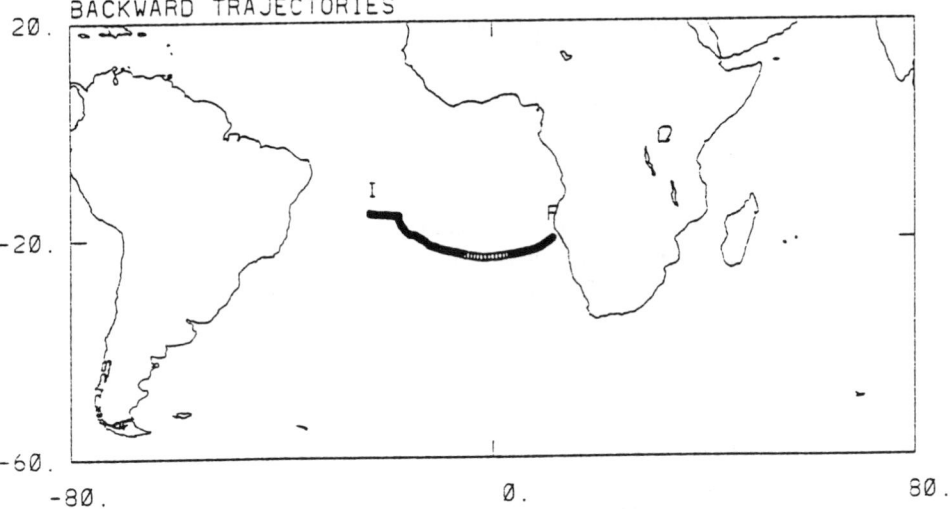

Figure 13.3: Example of backward trajectory calculation. The letters
T and 'F' represent the initial and final states, respectively.
Initial pressure 395.00 mb, final pressure 500 mb.

```
        program DIAGNOS
        parameter (l = 21,m = 13,np = 7 ,k = 1)
        real ug(l,m,np),vg(l,m,np),tg(l,m,np)
        real u(l,m,k),v(l,m,k),t(l,m,k)
        real zonav(m,k),areav (k),devzon(l,m,k),devaer(m,k)
        data gleft,gright,glat1,glat2/-25.,25.,-15.,15./
        data p ,del/500.,2.5/
c
c    Open input files.
c

        open (20,file='uv21.dat  ',status='old',readonly)
        open (30,file='temp21.dat',status='old',readonly)
c
c    Read from multilevel input data.
c
    15  format(10f8.2)
c
        do 13100 ip = 1, np
c
            read (20,15) ((ug(i,j,ip),i=1,l),j=1,m)
            read (20,15) ((vg(i,j,ip),i=1,l),j=1,m)
            read (30,15) ((tg(i,j,ip),i=1,l),j=1,m)
c
13100 continue
c
c    Select data for 500 mb level for this example.
c    Data in multilevel files are arranged as:
c    1000-850-700-500-300-200-100 mb
c
        do 13102 j  = 1 , m
        do 13102 i  = 1 , l
c
            u(i,j,k)   =  ug (i,j,4)
            v(i,j,k)   =  vg (i,j,4)
            t(i,j,k)   =  tg (i,j,4)
c
13102 continue
c
c    Compute the zonal and area average of the
c    temperature.
c
            call AVERAG1 (t,zonav,areav,l,m,k,gleft,gright,glat1,glat2.del)
```

```
c
c      Compute the deviation from the zonal average.
c
       call DEVIAT  (t,zonav,areav,devzon,devaer,l,m,k)
c
c      Compute the eddy available potential energy
c      and the zonal available potential energy.
c
       call APEZANDE (t,p,az,as,gleft,gright,glat1,glat2,del,l,m,k)
c
c      Compute the eddy kinetic  energy and the
c      zonal kinetic energy. Conversion of wind
c      speed inot cgs units is done first.
c
       do 13104 j  = 1,  m
       do 13104 i  = 1,  l
c
          u(i,j,1)  = u(i,j,k)* 100.
          v(i,j,1)  = v(i,j,k)* 100.
c
13104 continue

       call KEZANDE (u,v,ekz,eke,gleft,gright,glat1,glat2,del,l,m,k)
c
1000    format(21f8.2)
        stop
        end
c




        program TRAJECT
c
c      This program computes forward or backward trajectories
c      from/to any point for any u,v,w data set.
c
c      Definitions:
c
c      l        : number of points along east-west direction
c      m        : number of points north-south direction
c      n        : number of levels in the vertical direction
c
```

```
c        gl    : grid spacing
c        wl    : western most longitude
c        sl    : southern most latitude
c        gp    : interval between vertical levels
c                 (uniform thickness in this case)
c        nump  : total number of trajectories
c        nseg  : number of segments in each trajectory
c        kd    : number of days for trajectories
c        ttm   : total time in minutes
c        ttn   : interval in minutes between two observational
c                 data sets
c        nos   : number of sequential data sets
c        dt    : trajectory time step
c        is    : unit no of wind data of state of trajectories.
c        itd   : interval of trajectory data written.
c        td    : switch for forward or backward trajectories
c                 use td = 1 or -1 respectively
c

         parameter (l  =  17 ,  m =  13 , n = 10  ,gl = 2.5,gp = 100.)
         parameter (wl = -25., sl = -30.,bp = 100.,td = -1.,is = 41   )
         parameter (kd =   6 ,kob =   4 ,itd= 6    ,dt = 10.,nump = 1 )
         parameter(nseg = 721,ttm= 120.*60.,ttn = 6.*60.,nos = ttm/ttn+1)
         dimension u (l,m,n),v (l,m,n),w (l,m,n),dx(m),nq(nump)
         dimension up(l,m,n),vp(l,m,n),wp(l,m,n)
         dimension u1(l,m,n),v1(l,m,n),w1(l,m,n)
         dimension u2(l,m,n),v2(l,m,n),w2(l,m,n)
         dimension xs(nump,nseg),ys(nump,nseg),ps(nump,nseg)
c
c  Open input files .file trj.d contains trajectory output
c
         open (11,file='stp.d',status='old    ',form='formatted')
         open (13,file='trj.d',status='unknown',form='formatted')
c
c  The following files contain the 3 components of the wind
c  field for each 6 hourly interval and for the entire considered
c  domain.
c
         open(21,file='uvw.d000',status='old',form='formatted')
         open(22,file='uvw.d006',status='old',form='formatted')
         open(23,file='uvw.d012',status='old',form='formatted')
         open(24,file='uvw.d018',status='old',form='formatted')
         open(25,file='uvw.d024',status='old',form='formatted')
         open(26,file='uvw.d030',status='old',form='formatted')
```

```
      open(27,file='uvw.d036',status='old',form='formatted')
      open(28,file='uvw.d042',status='old',form='formatted')
      open(29,file='uvw.d048',status='old',form='formatted')
      open(30,file='uvw.d054',status='old',form='formatted')
      open(31,file='uvw.d060',status='old',form='formatted')
      open(32,file='uvw.d066',status='old',form='formatted')
      open(33,file='uvw.d072',status='old',form='formatted')
      open(34,file='uvw.d078',status='old',form='formatted')
      open(35,file='uvw.d084',status='old',form='formatted')
      open(36,file='uvw.d090',status='old',form='formatted')
      open(37,file='uvw.d096',status='old',form='formatted')
      open(38,file='uvw.d102',status='old',form='formatted')
      open(39,file='uvw.d108',status='old',form='formatted')
      open(40,file='uvw.d114',status='old',form='formatted')
      open(41,file='uvw.d120',status='old',form='formatted')
c
c     Compute the eastern, northern, and upper limits of the domain
c
      el            = wl+gl*(float(l-1))
      ul            = sl+gl*(float(m-1))
      ep            = bp+gp*(float(n-1))
c
c     Compute the x-grid spacing for all latitudes.
c
      call EPR (m,sl,gl,dx)
      dy            = gl*111.1*1000.
      do 13200 np = 1,nump
         nq(np)     = 1
13200 continue
      nsteps        = ttn/dt
c
c     Read data for terminal/starting point of trajectories
c     in degrees  of lat/long and pressure in mb.
c
      iu2           = is
      if (td.eq.1.) iu2 = 21
      tot           = 0.
      istp          = 0
c
c     Read the coordinates of the initial points.
c
      do 13202  i = 1, nump
      read (11,97) xs (i,1),ys (i,1),ps (i,1)
```

```
            write(6,97)  xs (i,1),ys (i,1),ps (i,1)
13202 continue
c
c     Read u,v and w data for all levels for initial position
c
            do 13204  k = 1, n
            read (iu2,92) ((u1(i,j,k),i=1,l),j=1,m)
            read (iu2,92) ((v1(i,j,k),i=1,l),j=1,m)
            read (iu2,92) ((w1(i,j,k),i=1,l),j=1,m)
13204 continue
c
c     Read u,v and w data for all levels for subsequent times
c
            nosr          = nos-1
            do 13216 knos  = 1, nosr
              iu2         = iu2+1*td
              write(14,*) iu2
            do 13206 k  = 1, n
            read (iu2,92) ((u2(i,j,k),i=1,l),j=1,m)
            read (iu2,92) ((v2(i,j,k),i=1,l),j=1,m)
            read (iu2,92) ((w2(i,j,k),i=1,l),j=1,m)
13206 continue
c
c     Computations for each trajectory step for the period
c     between two wind data sets.
c
      5  do 13208 nst = 1, nsteps
              tot          = tot +dt
              istp         = istp+1
c
c     Computation of u,v,w increments in one trajectory time step
c
            do 13210  k = 1, 10
            do 13210  j = 1, m
            do 13210  i = 1, l
              dudt        = ((u2(i,j,k)-u1(i,j,k))*(dt/ttn))
              dvdt        = ((v2(i,j,k)-v1(i,j,k))*(dt/ttn))
              dwdt        = ((w2(i,j,k)-w1(i,j,k))*(dt/ttn))
c
c     Computation of u,v,w at next trajectory time step
c
            u (i,j,k)     = u1(i,j,k)+dudt*(nst-1)
            v (i,j,k)     = v1(i,j,k)+dvdt*(nst-1)
```

```
      w (i,j,k)    = w1(i,j,k)+dwdt*(nst-1)
      up(i,j,k)    = u1(i,j,k)+dudt*(nst)
      vp(i,j,k)    = v1(i,j,k)+dvdt*(nst)
      wp(i,j,k)    = w1(i,j,k)+dwdt*(nst)
13210 continue
c
c     Computation for each individual trajectory
c
      do 13212 np = 1, nump
        sx1        = xs(np,istp)
        sy1        = ys(np,istp)
        sp1        = ps(np,istp)
      write(14,*) xs(np,istp),sx1,np,istp
      write(14,*) ys(np,istp),sy1,np,istp
      write(14,*) ps(np,istp),sp1,np,istp
      if(nq(np).eq.0) go to 8
c
c     For each trajectory for a step compute the terminal/
c     and lower pressure corner of grid cube and its distances
c     xd , yd and pd from this point and interpolate u,v and w
c     data at this point.
c
      call LOCATE (l,m,n,sx1,sy1,sp1,wl,sl,bp,gl,gp,
     &                   ir,jr,kr,xd,yd,pd,nump,np,nq)  .
      if (nq(np).eq.0) go to 8
      call INTRPL (u,l,m,n,ir,jr,kr,xd,yd,pd,up1)
      call INTRPL (v,l,m,n,ir,jr,kr,xd,yd,pd,vp1)
      call INTRPL (w,l,m,n,ir,jr,kr,xd,yd,pd,wp1)
c
c     Construct path to get next point at one time step difference
c     and to interpolate u,v,w data at that point and obtain mean
c     up,vp,wp for this segment
c
      call DISPL (up1,vp1,wp1,sx1,sy1,sp1,dx,dy
     &                   ,gl,m,jr,dt,td,sx2,sy2,sp2)
      call LOCATE (l,m,n,sx2,sy2,sp2,wl,sl,bp,gl,gp,
     &                   ir,jr,kr,xd,yd,pd,nump,np,nq)
      if(nq(np).eq.0) go to 8
      call INTRPL (up,l,m,n,ir,jr,kr,xd,yd,pd,up2)
      call INTRPL (vp,l,m,n,ir,jr,kr,xd,yd,pd,vp2)
      call INTRPL (wp,l,m,n,ir,jr,kr,xd,yd,pd,wp2)
c
      um           = (up1+up2)/2.0
```

```
            vm          = (vp1+vp2)/2.0
            wm          = (wp1+wp2)/2.0
c
c     Compute the path for one time step with mean u,v & w.
c     to get the position of this point and shift to it
c
            call DISPL (um,vm,wm,sx1,sy1,sp1,dx,dy,
      &                 gl,m,jr,dt,td,sx2,sy2,sp2)
c
c     house keeping work
c
            if(sx2.lt.wl.or.sx2.gt.el) nq(np) = 0
            if(sy2.lt.sl.or.sy2.gt.ul) nq(np) = 0
            if(sp2.lt.bp.or.sp2.gt.ep) nq(np) = 0
        8   if(nq(np).eq.1) go to 16
            sx2         = xs(np,istp)
            sy2         = ys(np,istp)
            sp2         = ps(np,istp)
       16   xs(np,istp+1)= sx2
            ys(np,istp+1)= sy2
            ps(np,istp+1)= sp2
        9   continue
13212 continue
13208 continue
            do 13214  k = 1, n
            do 13214  j = 1, m
            do 13214  i = 1, l
            u1(i,j,k)   = u2(i,j,k)
            v1(i,j,k)   = v2(i,j,k)
            w1(i,j,k)   = w2(i,j,k)
13214 continue
13216    continue
            istp        = istp+1
c
c     Writting the computed data for trajectories
c
            do 13218  i = 1, nump
               write (13,96) istp
            do 13218  k = 1, istp,itd
               kt       = (k-1)/6
            write(13,95) i,kt,ys(i,k),xs(i,k),ps(i,k)
13218 continue
       90   format(3f9.2)
```

```
92   format(6e13.6)
93   format(1x,6i7)
94   format(i5,3f10.3)
95   format(2i5,3e13.6)
96   format(3i7)
97   format(3f9.2)
     stop
     end
```

References

Anthes, R.A., 1977: A cumulus parameterization scheme utilizing a one-dimensional cloud model. *Mon. Wea. Rev.*, **105**, 270-286.

Arakawa, A., 1966: Computational design for long-term numerical integration of the equations of fluid motion: Two-dimensional incompressible flow. Part I. *Jour. Comp. Phys.*, **1**, 119-143.

Arakawa, A., 1971: A parameterization of cumulus convection and its application to numerical simulation of the tropical general circulation. Paper presented at the 7th Tech. Conf. on Hurricanes and Tropical Meteorology, Barbados, A.M.S.

Arakawa, A. and W.H. Schubert, 1974: Interaction of a cumulus cloud ensemble with the large-scale environment. Part I. *Jour. Atmos. Sci.*, **31**, 674-701.

Arkin, P.A., 1994: The Global Precipitation Climatology Project: First Algorithm Intercomparison Project. *Bulletin of the American Meteorological Society*, **75**, 401-419.

Barnes, S.L., 1964: A technique for maximizing details in numerical weather map analysis. *J. Appl. Meteor.*, **3**, 396-409.

Bergman, K.H., 1979: Multivariate analysis of temperature and winds using optimum interpolation. *Mon. Wea. Rev.*, **107**, 1423- 1444.

Bergthorsson, P. and B.R. Döös, 1955: Numerical weather map analysis. *Tellus*, **7**, 329-340.

Beven, J., 1994: Tropical cyclone-environmental interactions during recurvature: an observational and modeling study. Ph.D. dissertation, Florida State University, Tallahassee.

Blackadar, A.K., 1962: The vertical distribution of wind and turbulent exchange in a neutral atmosphere. *J. Geophys. Res.*, **67**, 3095-3102.

Bounoua, L. and T.N. Krishnamurti, 1993: Influence of soil moisture on the Sahelian climate prediction (Part I). *Meterol. Atmos. Phys.*, **52**, 183-203.

Businger, J.A., J.C. Wyngaard, Y. Izumi, and E.F. Bradley, 1971: Flux profile relationship in the atmospheric surface layer. *Jour. Atmos. Sci.,* 28, 181-189.

Chang, C.B., 1978: On radiative interactions in an African disturbance. Ph.D. dissertation, Florida State University, Tallahassee, 1-163. Available from University Microfilms, University of Michigan, Ann Arbor.

Charnock, H., 1955: Wind stress on a water surface. *Quart. J. Roy. Met. Soc.,* 81, 639-640.

Cressman, G., 1959: An operational objective analysis system. *Mon. Wea. Rev.,* 87, 367-374.

Curran, P.J., 1980: Multispectral remote sensing of vegetation amount. *Progr. Phys. Geog.,* 4, 175-184.

Deardorff, J.W., 1972: Numerical investigation of neutral and unstable planetary boundary layers. *Jour. Atmos. Sci.,* 29, 91-115.

Delsol, F., K. Miyakoda, and R. Clarke, 1971: Parameterized processes in the surface boundary layer of an atmospheric circulation model. *Quart. J. Roy. Met. Soc.,* 97, 181-208.

Dey, C.H. and L.L. Morone, 1985: Evolution of the National Meteorological Center global data assimilation system: January 1982 - December 1983. *Mon. Wea. Rev.,* 113, 304-318.

DiMego, G.J., 1988: The National Meteorological Center regional analysis system. *Mon. Wea. Rev.,* 116, 977-1000.

Fels and Schwarzkopf, 1981: An efficient, accurate algorithm for calculating CO_2 15 micron band cooling rates. *J. Geophys. Res.,* 86, 1205-1232.

Filiberti, M.A., L. Eymard, and B. Urban, 1993: Assimilation of satellite precipitable water in a meteorological forecast model. *Mon. Wea. Rev.,* 122, 304-318.

Gandin, L.S., 1963: Objective analysis of meteorological fields. Translated from Russian, Israel Program for Scientific Translations, Jerusalem, 242 pp. (NTIS TT-65-50007).

Gurney, R.J., J.L. Foster, and C.L. Parkinson, 1993: Atlas of satellite observations related to global change. Cambridge University Press, Cambridge, 470.

Halpern, D. and W. Knauss, 1993: An Atlas of Monthly Mean Distributions of SSMI Surface Wind Speed, ARGOS Buoy Drift, AVHRR/2 Sea Surface Temperature, and ECMWF Surface Wind Components During 1991. Jet Propulsion Laboratory Publication 93-10.

Haltinger, G.J. and R.T. Williams, 1979: *Numerical prediction and dynamic meteorology*. Wiley Publication, New York, pp. 1-477.

Holton, J.R., 1992: *An Introduction to Dynamic Meteorology*, Academic Press, New York, 511 pp.

Houghton, Henry G., 1985: *Physical Meteorology*, The MIT Press, Boston, 442 pp.

Joseph, J.H., 1966: Calculation of radiative heating in numerical general circulation models. Tech. Rep. No. 1, Dept. of Meteorology, University of California, Los Angeles, 60 pp.

Kanamitsu, M., 1975: On numerical prediction over a global tropical belt. Report No. 75-1, Dept. of Meteorology, Florida State University, Tallahassee, pp. 1-282.

Katayama, A., 1972: A simplified scheme for computing radiative transfer in the troposphere. Tech. Rep. No. 6, Dept. of Meteorology, University of California, Los Angeles, 77 pp.

Kondratyev, K.Y., 1972: Radiation processes in the atmosphere. World Meteorological Organization, WMO-No. 309, Geneva, Switzerland, 214 pp.

Krishnamurti, T.N., 1968: A diagnostic balance model for studies of weather systems of high and low latitudes, Rossby number < 1. *Mon. Wea. Rev.*, **96**, 197-207.

Krishnamurti, T.N., 1974: Lectures on tropical meteorology in the dynamics of the tropical atmosphere. Published as colloquium notes. National Center for Atmospheric Research, Boulder, CO 105 pp.

Krishnamurti, T.N., H.L. Pan, C.B. Chang, J. Ploshay, D. Walker, and A.W. Oodally, 1979: Numerical Weather Prediction for Gate. *Quart. J. Roy. Met. Soc.*, **105**, 979-1010.

Krishnamurti, T.N., Y. Ramanathan, H.L. Pan, R.J. Pasch, and J. Molinari, 1980: Cumulus parameterization and rainfall rates I. *Mon. Wea. Rev.*, **108**, 465-472.

Krishnamurti, T.N., S. Low-Nam, and R. Pasch, 1983: Cumulus parameterization and rainfall rates II. *Mon. Wea. Rev.*, **111**, 815-828.

Krishnamurti, T.N., A. Kumar, and X. Li, 1986: Results of extensive integrations with simple NWP models over the entire tropics during FGGE. *Tellus*, **39A**, 152-160.

Kumar, A., 1990: Generalized dynamic normal mode initialization. Ph.D. thesis, Dept. of Meteorology, Florida State University, Tallahassee.

Kuo, H.L., 1965: On formation and intensification of tropical cyclones through latent heat release by cumulus convection. *Jour. Atmos. Sci.*, **22**, 40-63.

Kuo, H.L., 1974: Further studies of the parameterization of the influence of cumulus convection on large-scale flow. *Jour. Atmos. Sci.*, **32**, 1232-1240.

Lenschow, D.H., 1970: Airplane measurements of the planetary boundary structure. *J. Appl. Met.*, **9**, 874-884.

Lord, S.J., 1978: Development and observational verification of a cumulus cloud parameterization. Ph.D. thesis, UCLA.

Lord, S.J., 1982: Interactions of a cumulus cloud ensemble with the large-scale environment III. Semi-prognostic test of the Arakawa-Schubert theory. *Jour. Atmos. Sci.*, **39**, 88-103.

Lorenz, E.N., 1967: The nature and theory of the general circulation of the atmosphere. World Meteorological Organization, WMO-No. 218. Geneva, Switzerland, 161 pp.

Louis, J.F., 1979: A parametric model of the vertical eddy fluxes in the atmosphere. *Bound. Layer. Meteor.*, **17**, 187-202.

Mathur, M.B., 1970: A note on a improved quasi-Lagrangian advective scheme for primitive equations. *Mon. Wea. Rev.*, **98**, 214-219.

Nickerson, E., 1965: A numerical study in buoyant convection involving the use of a heat source. *Jour. Atmos. Sci.*, **22**, 412-418.

Nyhoff, L. and S. Leestma, 1988: *Fortran 77 for Engineers and Scientists*, Macmillan Publishing Company, New York, 590 pp.

Olson, W.S., F.S. Fontaine, W.L. Smith, and R.H. Achtor, 1990: Recommended algorithms for the retrieval of rainfall rates in the tropics using SSM/I (DMSP-F8). Manuscript, University of Wisconsin, Madison, 10 pp.

Paegle, J., 1966: Computation of three-dimensional trajectories. Dept. of Meteorology, UCLA.

Panofsky, J., 1949: Objective weather-map analysis. *Journal of Meteorology*, **6**, 386-392.

Petterssen, S., 1956: *Weather analysis and forecasting*, I, Chapter 9. McGraw-Hill Book Company, New York.

Phillips, N.A., 1963: Geostrophic motion. *Rev. Geophysics*, **1**, 123-176.

Posey, J.W. and P.F. Clapp, 1964: Global distribution of normal surface albedo. *Geofisica Internacional* (Mexico), **4**, 33-48.

Rodgers, C.D., 1967: The use of emissivity in atmospheric radiation calculation. *Quart. J. Roy. Met. Soc.*, **93**, 43-45.

Sellers, P.J., 1985: Canopy reflectance, photosynthesis and transpiration. *Int. J. Remote. Sensing*, **6**, 1335-1372.

Slingo, J.M., 1987: The development and verification of a cloud prediction scheme for the ECMWF model. *Q. J. Roy. Met.Soc.*, **113**, 899-927.

Smeda, M.S., 1977: Incorporation of planetary boundary layer processes into numerical forecasting models. Report No. DM-23, Dept. of Meteorology, University of Stockholm, 1-45.

Staley and Jurice, 1970: Flux emissivity tables for water vapor, carbon dioxide, and ozone. *J. Appl. Meteo.*, **9**, 365-372.

Stephens, J.J. and K.W. Johnson, 1978: Rotational and divergent wind potentials. *Mon. Wea. Rev.*, **106**, 1452-1457.

Sugi, M., 1986: Dynamic normal mode initialization. *J. Meteorol. Soc. Japan*, **64**, 623-636.

Sutcliffe, R.C., 1947: A contribution to the problem of development. *Quart. J. Roy. Met. Soc.*, **73**, 370-383.

Thiebaux, H.J. and M.A. Pedder, 1987: *Spatial Objective Analysis with Applications in Atmospheric Science.* Academic Press, Orlando, FL, 299 pp.

Tripoli, G.J. and T.N. Krishnamurti, 1975: Low-level flows over the GATE area during summer 1972. *Mon. Wea. Rev.*, **103**, 197-216.

Tucker, C.J. and L.D. Miller, 1977: Contribution of the soil spectra to grass canopy spectral reflectance. *Photogrammetric Engineering and Remote Sensing*, **43**, 721-726.

Tucker, C.J., B.N. Holben, J.H. Elgin and J.E. McMurtrey, 1981: Remote sensing of total dry-matter accumulation in winter wheat. *Remote Sensing of Environment*, **11**, 171-189.

Tucker, C.J. and P.J. Sellers, 1986: Satellite remote sensing of primary production. *Int. J. of Remote Sensing*, **7**, 1395-1416.

Wallace, J.M. and P.V. Hobbs, 1977: *Atmospheric Science: an Introductory Survey*, Academic Press, New York, 467 pp.

Xue, J., 1990: Personal communication. Florida State University, Tallahassee.

Yanai, M., S. Esbensen, and J.H. Chu, 1973: Determination of bulk properties of tropical cloud clusters from large-scale heat and moisture budgets. *Jour. Atmos. Sci.*, **30**, 611-627.

List of Subroutines

Subroutine	Function
DDX2	Performs the second order finite difference.
DDX4	Performs the fourth order finite difference. The computation is cyclic.
JAC	Performs the Arakawa Jacobian.
LAP94	Performs the iterated nine points fourth order Laplacian.
LAP92	Performs the second order Laplacian using a nine-point stencil.
LAP52	Performs the second order Laplacian using a five-point stencil.
PLNSFC	Solves a system of linear equations.
RELAXW	Computes vertical velocity via relaxation.
SIGMAL	Computes the stability change due to heat release.
LAP	General Computation of the Laplacian of any 3-D function.
MYMINMAX	Searches for the minimum and maximum of a 2-D field.
KINOMGA	Evaluates vertical velocity using kinematic method.
VMOTION	Calculates the vertical motion. Correction is applied.
VINTGRL	Computes the vertical integral.
ROME	Solves a three-dimensional Poisson equation.
RELAX	Solves the Poisson equation using relaxation.
PSICHI	Computes streamfunction and velocity potential.
FOURT	Performs the fast Fourier transform (FFT).
ZFIELD	Computes geopotential height from streamfunction using balance models.
ZERO	Initializes fields to zero.
OBJAN	Performs an objective analysis of the wind field using the successive corrections method.
SMOOTH	Performs smoothing on a 2-D array.
BUFF	Reads/Writes files to/from specified units.
OBJAN2	Performs an objective analysis following Barnes' method.
INTERP	Computes the weights for Barnes' scheme.
QFRMRH	Returns specific humidity values from an input of relative humidity.
RHFRMTD	Returns relative humidity values from an input of dew point temperature.

QFRMTD	Returns specific humidity values from an input of dew point temperature.
TDFRMQ	Returns dew point values from an input of specific humidity.
TDFRMRH	Returns dew point values from an input of relative humidity.
RHFRMQ	Returns relative humidity values from an input of specific humidity.
LCL	Finds the level of condensation (LCL).
TRIDI	Solves a tridiagonal matrix system.
FIVEDI	Solves a pentadiagonal matrix system.
THETAE	Computes a vertical profile of equivalent potential temperature.
MOIST	Constructs a moist adiabat.
ADJTOP	Determines the top of moist convective adjustment.
DCONDADJ	Performs the dry convective adjustment.
BASIC	Defines the initial state for Nickerson's cloud model.
RELAX1	Solves Poisson equation.
CLOUD	Writes outputs.
LAPLAC	Calculates the Laplacian.
CVHEAT	Computes the heating and moistening contributions due to deep convection a la Kuo (1965).
STBHEAT	Computes the large scale condensation heating.
BLKFLX	Computes surface heat fluxes using bulk aerodynamic.
HUMSPC	Computes the specific humidity as function of dew point (function).
WETCNS	Computes latent heat of evaporation with respect to water and ice.
FLXSRF	Sets the parameters necessary to the computation of the surface fluxes.
TG	Uses a coupled surface energy balance to solve for surface temperature.
SFLX	Returns surface sensible heat, latent heat and momentum fluxes.
SFXPAR	Computes surface parameters for the surface fluxes calculations.
CONRAD	Computes the declination angle.
RAD	Computes the shortwave and longwave radiative fluxes.
SLR	Performs the computation of shortwave radiation.
RLW	Performs the calculation of longwave radiation.
EMTAB	Returns the emissivity values as function of the optical depth in logarithmic scale.

BOUND	Used to obtain southern and northern boundary values.
CONST	Defines constants and parameters required by program INFIELD.
CYCLE	Used to create cyclic boundary in the zonal direction.
JACMOD	Computes the Jacobian for a cyclic domain.
LAPMOD	Computes the Laplacian for a cyclic domain.
RELAXMOD	Solves the Poisson's equation over a cyclic domain.
STREAMF	Computes the streamfunction field.
ZFIELDMOD	Computes the height field from the streamfunctions.
TERR	Computes the gradient of the terrain field.
ENERGY	Calculates the domain average of energy terms.
EQUAL	Equalizes two arrays.
INIT	Defines initial parameters for the barotropic model.
LARGE	Searches for the largest value of a 3-D field.
RELAXT	Solves the Poisson equation with known boundary values.
SMALL	Searches for the smallest value of a 3-D field.
VORT	Computes the absolute vorticity.
CONST2	Defines the constants for the single level model.
FCST	Performs the time integration of the single level model.
INDATA	Reads the input data.
INTERP2	Performs the semi-Lagrangian advection computation.
INVART	Computes the closed domain invariants of the single level primitive equation model.
ZEROS	Initializes the values of the variables to zero.
KEZANDE	Computes the zonal kinetic energy and the eddy kinetic energy.
APEZANDE	Computes the eddy available potential and the zonal available potential energy.
STASTB	Computes the static stability parameter.
AVMERID	Computes the meridional mean of a 2-D field.
INTEGR	Performs vertical integration.
AVERAG1	Computes the zonal average of 'a' as well as its area average.
DEVIAT	Computes the departure of 3-D field 'a' from the zonal average and the departure of the zonal average of 'a' from the area average.
CECZTERM	Computes the conversion of eddy available potential energy to eddy kinetic energy. Also computes the conversion of the zonal available potential energy to zonal kinetic energy.

DERIVE	Computes the derivative of a 2-D field 'a' with respect to y (the meridional coordinate).
PRESDE	Computes the derivative of a 2-D field 'a' with respect to pressure.
CKCATERM	Computes the conversion of zonal kinetic energy to eddy kinetic energy. Also computes the conversion of zonal available energy to eddy available potential energy.
GZGETERM	Computes the energy generation terms.
LOCATE	Finds the i, j, k indices of the southwest and lower pressure corner of the grid box.
INTRPL	Performs a trilinear interpolation of the wind field components.
DISPL	Computes the displacement along the trajectory in one time step.
EPR	Computes the grid spacing along the x-axis for all latitudes.

Index